Studies in Computational Intelligence   470

Editor-in-Chief

Prof. Janusz Kacprzyk
Systems Research Institute
Polish Academy of Sciences
ul. Newelska 6
01-447 Warsaw
Poland
E-mail: kacprzyk@ibspan.waw.pl

For further volumes:
http://www.springer.com/series/7092

Stefka Fidanova (Ed.)

Recent Advances in Computational Optimization

 Springer

Editor
Prof. Stefka Fidanova
Institute of Information and Communication Technology
Bulgarian Academy of Sciences
Sofia
Bulgaria

ISSN 1860-949X   ISSN 1860-9503   (electronic)
ISBN 978-3-319-03307-5   ISBN 978-3-319-00410-5 (eBook)
DOI 10.1007/978-3-319-00410-5
Springer Cham Heidelberg New York Dordrecht London

Preface

Every day we solve optimization problems. Optimization occurs in the minimizing time and cost or the maximization of the profit, quality and efficiency. Many real world problems arising in engineering, economics, medicine and other domains can be formulated as optimization tasks. Such problems are frequently characterized by non-convex, non-differentiable, discontinuous, noisy or dynamic objective functions and constraints which ask for adequate computational methods.

This volume is a result of very vivid and fruitful discussions held during the Workshop on Computational Optimization. The participants have agreed that the relevance of the conference topic and quality of the contributions have clearly suggested that a more comprehensive collection of extended contributions devoted to the area would be very welcome and would certainly contribute to a wider exposure and proliferation of the field and ideas.

The volume include important real problems like parameter settings for controlling processes in bioreactor, robot skin wiring, strip packing, project scheduling, tuning of PID controller and so on. Some of them can be solved applying traditional numerical methods, but others needs huge amount of computational resources. Therefore for them are more appropriate to develop an algorithms based on some metaheuristic method like evolutionary computation, ant colony optimization, constrain programming etc.

April 2013

Stefka Fidanova
Co-Chair
WCO'2012

Organization

Workshop on Computational Optimization (WCO 2012) is organized in the framework of FEDERATED CONFERENCE ON COMPUTER SCIENCE AND INFORMATION SYSTEMS FedCSIS - 2012

Conference Co-chairs

Stefka Fidanova	IICT (Bulgarian Academy of Sciences, Bulgaria)
Antonio Mucherino	IRISA (Rennes, France)
Josef Tvrdik	University of Ostrava (Czech Republic)
Daniela Zaharie	West University of Timisoara (Romania)

Program Committee

Andonov, Rumen	IRISA and University of Rennes 1, Rennes, France
Bartl, David	University of Ostrava, Czech Republic
Brest, Janez	University of Maribor, Slovenia
Hoai An, Le Thi	University of Lorraine, France
Hosobe, Hiroshi	National Institute of Informatics, Japan
Iiduka, Hideaki	Kyushu Institute of Technology, Japan
Judice, Joaquim	Instituto Telecomunicaes, Portugal
Kukal, Jaromir	Czech Technical University in Prague, Czech Republic
Lampinen, Jouni	University of Vaasa, Finland
Lavor, Carlile	IMECC-UNICAMP, Campinas, Brazil
Marinov, Pencho	Bulgarian Academy of Science, Bulgaria
Mesjasz, Mariusz	WUT, Poland
Neri, Ferrante	University of Jyvskyl, Finland
Pardalos, Panos	University of Florida, United States
Penev, Kalin	Southampton Solent University, United Kingdom
Siarry, Patrick	Universite Paris XII Val de Marne, France

Stefanov, Stefan	Neofit Rilski University, Bulgaria
Stuetzle, Tomas	Universite Libre de Bruxelles, Belgium
Suganthan, Ponnuthurai	Nanyang Technological University, Singapore
Vrahatis, Michael	University of Patras, Greece
Zilinskas, Antanas	Research Institute of Mathematics and Informatics, Lithuania

Contents

Intuitionistic Fuzzy Logic as a Tool for Quality Assessment of Genetic Algorithms Performances

Maria Angelova, Krassimir Atanassov, and Tania Pencheva

Institute of Biophysics and Biomedical Engineering
Bulgarian Academy of Sciences
105 Acad. G. Bonchev Str., 1113 Sofia, Bulgaria
{maria.angelova,tania.pencheva}@biomed.bas.bg, krat@bas.bg

Abstract. Intuitionistic fuzzy logic (IFL) has been implemented in this investigation aiming to derive intuitionistic fuzzy estimations of *S. cerevisiae* fed-batch cultivation model parameters obtained using standard simple (SGA) and multi-population (MpGA) genetic algorithms. Performances of MpGA have been tested before and after the application of the procedure for purposeful model parameters genesis at three different values of generation gap, proven as the most sensitive genetic algorithms parameter toward convergence time. Results obtained after the implementation of intuitionistic fuzzy logic for MpGA performances assessment have been compared and MpGA at GGAP = 0.1 after the purposeful model parameters genesis procedure application has been distinguished as the fastest and the most reliable one. Further, the prominent MpGA at GGAP = 0.1 has been compared to SGA at GGAP = 0.1. Obtained results have been assessed applying IFL and the most reliable algorithm has been distinguished.

1 Introduction

Among a number of searching tools, genetic algorithms (GA) are one of the methods based on biological evolution, inspired by Darwins theory of survival of the fittest. GA [1] are directed random search techniques, based on the mechanics of natural selection and genetics, and seek for the global optimal solution in complex multidimensional search space by simultaneously evaluating many points in the parameter space. Some properties such as hard problems solving, noise tolerance, easiness to interface and hybridize, make GA a suitable and quite workable tool especially for tasks which are not completely determined. Such an intractable problem and a real challenge for researchers is the parameter identification of fermentation processes models [2,3,4,5,6]. Modeling of fermentation processes, known as complex, dynamic systems with interdependent and time-varying process variables, is a specific task, rather difficult to be solved. Failure of conventional optimization methods to reach to a satisfactory solution for parameters identification of fermentation process model [2,5] provokes idea genetic algorithms to be tested as an alternative technique.

S. Fidanova (Ed.): *Recent Advances in Computational Optimization*, SCI 470, pp. 1–13.
DOI: 10.1007/978-3-319-00410-5_1   © Springer International Publishing Switzerland 2013

Goldberg [1] initially presents the standard single-population genetic algorithm (SGA) inspired by natural genetics. SGA searches a global optimal solution using three main genetic operators in a sequence selection, crossover and mutation. More similar to nature is multi-population genetic algorithm (MpGA), since there many populations, called subpopulations, evolve independently from each other. After a certain number of generations a part of individuals are distributed between the subpopulations (migration).

According to [1,7] structure of standard SGA could be shortly presented below in eight steps:

Begin
1. [**Start**] Generate random population of n chromosomes
2. [**Objective function**] Evaluate the objective function of each chromosome x in the population
3. [**Fitness function**] Find the fitness function of each chromosome x in the population
4. [**New population**] Create a new population by repeating following steps:
4.1. [**Selection**] Select parent chromosomes from the population according to their fitness function
4.2. [**Crossover**] Cross over the parents to form new offspring with a crossover probability
4.3. [**Mutation**] Mutate new offspring at each locus with a mutation probability
5. [**Accepting**] Place new offspring in a new population
6. [**Replace**] Use new generated population for a further run of the algorithm
7. [**Test**] If the end condition is satisfied, stop and return the best solution in current population, else move to **Loop** step
8. [**Loop**] Go to **Fitness step**.
End

In the case of MpGA the algorithm starts not with random population of n chromosomes but with the generation of k random subpopulations each of them with n chromosomes. After that all the steps are performed not for the populations but for the subpopulations. Additionally, a new step appears after Step 6 **Replace**, namely the step of

[**Migration**] Migration of individuals between the subpopulations after following isolation time.

When GA are applied for the purposes of model parameter identification, there are many operators, functions, parameters and settings that may vary depending on the considered problems [1,8]. In [8] generation gap, which is the fraction of the population to be reproduced, has been investigated altogether with crossover and mutation rate towards convergence time. Among them three, generation gap (GGAP) has been distinguished as the most sensitive genetic algorithm parameter. Up to almost 40% of the algorithm calculation time can be saved in the case of MpGA application using GGAP = 0.5 instead of 0.9 without loss of model accuracy. The same outcome has been achieved when SGA is applied. Obtained promising results in [8] provoke the idea of subsequent reduction of the generation gap value. Thus the topic of the present work is to be investigated the MpGA quality of performance for three different values of generation gap GGAP = 0.9, GGAP = 0.5 and GGAP = 0.1. Additionally, the performance quality of SGA and MpGA at GGAP = 0.1 to be compared based on IFL.

The quality of GA performance could be appraised by some representative criteria such as the objective function value and the algorithm convergence time As an alternative for assessing the quality of different algorithms intuitionistic fuzzy logic (IFL) might be applied for various purposes. In order to construct the degree of validity and non-validity it is required the algorithms to be performed in two different intervals of model parameters variation. One interval could be determined as so-called broad range known from the literature [6]. The other one, called narrow range, is user-defined and might be obtained using different criteria e.g. based on the minimum and maximum values, or on the average ones, or after the implementation of the procedure for purposeful model parameters genesis [9].

The aim of this study is intuitionistic fuzzy estimations to be applied for assessing the multi-population genetic algorithm at three different values of generation gap. After that performance of standard SGA towards standard MpGA to be compared at the most reliable GGAP value when GA have been implemented in parameter identification of *S. cerevisiae* fed-batch cultivation.

Aiming to save decreased convergence time while keeping or even improving model accuracy, intuitionistic fuzzy estimations overbuild the results obtained after procedure of purposeful model parameters genesis.

2   Background

2.1   Intuitionistic Fuzzy Estimations

In intuitionistic fuzzy logic (IFL) [10,11] if p is a variable then its truth-value is represented by the ordered couple

$$V(p) = \langle M(p), N(p) \rangle \tag{1}$$

so that $M(p), N(p), M(p) + N(p) \in [0, 1]$, where $M(p)$ and $N(p)$ are degrees of validity and of non-validity of p. These values can be obtained applying different formula depending on the problem considered.

For the purpose of this investigation the degrees of validity/non-validity can be obtained, e.g., by the following formula:

$$M(p) = \frac{m}{u}, \quad N(p) = 1 - \frac{n}{u}, \tag{2}$$

where m is the lower boundary of the narrow range; u - the upper boundary of the broad range; n - the upper boundary of the narrow range.

If there is a database collected having elements with the form $< p, M(p), N(p) >$, different new values for the variables can be obtained. In case of two records in the database, the new values might be as follows:

$$V_{opt} = \langle \max(M_1(p), M_2(p)), \min(N_1(p), N_2(p)) \rangle, \tag{3}$$

$$V_{aver} = \left\langle \frac{M_1(p) + M_2(p)}{2}, \frac{N_1(p) + N_2(p)}{2} \right\rangle, \tag{4}$$

$$V_{pes} = \langle \min(M_1(p), M_2(p)), \max(N_1(p), N_2(p)) \rangle, \tag{5}$$

Therefore, for each p

$$V_{pes}(p) \le V_{aver}(p) \le V_{opt}(p).$$

In case of three records in the database, the following new values can be obtained:

$$V_{strong\_opt} = \langle M_1(p) + M_2(p) + M_3(p) - M_1(p)M_2(p) - M_1(p)M_3(p) -$$

$$- M_2(p)M_3(p) + M_1(p)M_2(p)M_3(p), N_1(p)N_2(p)N_3(p) \rangle, \tag{6}$$

$$V_{opt} = \langle \max(M_1(p), M_2(p), M_3(p)), \min(N_1(p), N_2(p), N_3(p)) \rangle, \tag{7}$$

$$V_{aver} = \left\langle \frac{M_1(p) + M_2(p) + M_3(p)}{3}, \frac{N_1(p) + N_2(p) + N_3(p)}{3} \right\rangle, \tag{8}$$

$$V_{pes} = \langle \min(M_1(p), M_2(p), M_3(p)), \max(N_1(p), N_2(p), N_3(p)) \rangle, \tag{9}$$

$$V_{strong\_pes} = \langle M_1(p)M_2(p)M_3(p), N_1(p) + N_2(p) + N_3(p) - N_1(p)N_2(p)$$

$$- N_1(p)N_3(p) - N_2(p)N_3(p) + N_1(p)N_2(p)N_3(p) \rangle \tag{10}$$

Therefore, for each p

$$V_{strong\_pes}(p) \le V_{pes}(p) \le V_{aver}(p) \le V_{opt}(p) \le V_{strong\_opt}(p).$$

2.2 Procedure for Purposeful Model Parameter Genesis

The procedure for purposeful model parameter genesis (PMPG) has been originally developed and firstly applied for simple genetic algorithms [9]. Aiming to obtain reliable results in parameter identification of a fermentation process model when using genetic algorithms, a great number of algorithm runs have to be executed because of their stochastic nature. Firstly, the genetic algorithm searches for solutions of model parameters in wide but reasonably chosen boundaries according to the statements in [6]. When results from many algorithms executions were accumulated and analyzed, they showed that the values of model parameters can be assembled and predefined boundaries of model parameters could be restricted. That provoked the idea for PMPG, which results in the defining of more appropriate boundaries for variation of the model parameters values. The procedure application leads to decrease convergence time while at least saving or even improving the model accuracy.

2.3   Procedure for Genetic Algorithms Quality Assessment Applying IFL

As mentioned above, the implementation of IFL to assess the quality of genetic algorithms requires constructing the degree of validity and non-validity in two different intervals of model parameters variation: so-called broad range as known from the literature and so-called narrow range which is user-defined. A procedure for assessment of algorithm quality performance (AAQP) is proposed [12] to evaluate the quality of genetic algorithms applying IFL. At the beginning, a number of runs of each of the algorithms, object of the investigation, have to be performed in both broad and narrow ranges of model parameters. Then the average values of the objective function, algorithms convergence time and each of the model parameters for each one of the ranges and each one of the investigated algorithms are determined. Next the degrees of validity/non-validity for each of the algorithms, object of the investigation, are determined according to (2). Then, according to (3)-(5) in case of two objects considered, *optimistic, average* and *pessimistic* values are defined; or, according to (6)-(10) in case of three objects, *strong optimistic, optimistic, average, pessimistic* and *strong pessimistic* values are calculated for each one of the model parameters. Next determined in such way values are assigned to each of the model parameters for each of the ranges for each of the algorithms. Finally, based on these assigns, the quality of each one of considered algorithm is assessed.

2.4   Mathematical Model of *S. cerevisiae* Fed-Batch Cultivation

Experimental data of *S. cerevisiae* fed-batch cultivation is obtained in the *Institute of Technical Chemistry - University of Hannover*, Germany [2]. The cultivation of the yeast *S. cerevisiae* is performed in a 2 l reactor, using a Schatzmann medium. Glucose in feeding solution is 35 g/l. The temperature was controlled at 30°C, the pH at 5.5. The stirrer speed was set to 1200 rpm. Biomass and ethanol were measured off-line, while substrate (glucose) and dissolved oxygen were measured on-line.

Mathematical model of *S. cerevisiae* fed-batch cultivation is commonly described as follows, according to the mass balance [2]:

$$\frac{dX}{dt} = \mu X - \frac{F}{V} X \tag{11}$$

$$\frac{dS}{dt} = -q_S X + \frac{F}{V}(S_{in} - S) \tag{12}$$

$$\frac{dE}{dt} = q_E X - \frac{F}{V} E \tag{13}$$

$$dt = -q_{O_2} X + k_L^{O_2} a (O_2^* - O_2) \tag{14}$$

$$\frac{dV}{dt} = F \tag{15}$$

where X is the concentration of biomass, [g/l]; S - concentration of substrate (glucose), [g/l]; E - concentration of ethanol, [g/l]; O_2 - concentration of oxygen, [%]; O_2^* - dissolved oxygen saturation concentration, [%]; F - feeding rate, [l/h]; V - volume of bioreactor, [l]; $k_L^{O_2}a$ - volumetric oxygen transfer coefficient, [1/h]; S_{in} - glucose concentration in the feeding solution, [g/l]; μ, q_S, q_E, q_{O_2} - specific growth/utilization rates of biomass, substrate, ethanol and dissolved oxygen, [1/h]. All functions are continuous and differentiable.

The fed-batch cultivation of *S. cerevisiae* considered here is characterized by keeping glucose concentration equal to or below to its critical level ($S_{crit} = 0.05$ g/l), sufficient dissolved oxygen in the broth $O_2 \geq O_{2crit}$ ($O_{2crit} = 18\%$) and availability of ethanol in the broth. This state corresponds to the so called mixed oxidative state (FS II) according to functional state modeling approach [2]. Hence, specific rates in Eqs. (11)-(15) are:

$$\mu = \mu_{2S}\frac{S}{S+k_S} + \mu_{2E}\frac{E}{E+k_E}, \quad q_S = \frac{\mu_{2S}}{Y_{SX}}\frac{S}{S+k_S}$$

$$q_E = -\frac{\mu_{2E}}{Y_{EX}}\frac{E}{E+k_E}, \quad q_{O_2} = q_E Y_{OE} + q_S Y_{OS} \tag{16}$$

where μ_{2S}, μ_{2E} are the maximum growth rates of substrate and ethanol, [1/h]; k_S, k_E - saturation constants of substrate and ethanol, [g/l]; Y_{ij} - yield coefficients, [g/g]; and all model parameters fulfill the non-zero division requirement.

As an optimization criterion, mean square deviation between the model output and the experimental data obtained during cultivation has been used:

$$J_Y = \sum(Y - Y^*)^2 \rightarrow min, \tag{17}$$

where Y is the experimental data, Y^* - model predicted data, $Y = [X, S, E, O_2]$.

3   MpGA Quality Assessment at Different Values of GGAP

The procedure for purposeful model genesis has been applied to parameter identification of *S. cerevisiae* fed-batch cultivation using MpGA. Following model (11)-(16) of *S. cerevisiae* fed-batch cultivation, nine model parameters have been estimated altogether, applying MpGA with three different GGAP values, proven as the most sensitive genetic algorithms parameter towards the algorithms convergence time [8]. The values of other GA parameters and type of genetic operators in MpGA considered here are tuned according to [8]. GA is terminated when a certain number of generations is fulfilled, in this case 100. Scalar relative error tolerance *RelTol* is set to 1e-4, while the vector of absolute error tolerances (all components) *AbsTol* - to 1e-5. Parameter identification of the model (11)-(16) has been performed using *Genetic Algorithm Toolbox* [13] in *Matlab 7* environment. All the computations are performed using a PC Intel Pentium 4 (2.4 GHz) platform running Windows XP.

The quality of MpGA performance is assessed before and after application of PMPG, that means that the narrow range is obtained applying PMPG. The obtained results are firstly analyzed according to achieved objective function values and convergence time. For each GGAP value the minimum and the maximum of the objective function are determined, and the levels of performance according to PMPG [9] have been constructed. According to the values of obtained objective function there are only two levels of performance in MpGA. The best results hit the interval $[minJ; minJ + \Delta - \varepsilon]$ and they form the top level of MpGA performance. The worse solutions for the objective function fall in the interval $[minJ + \Delta; maxJ]$ and thus create the low level of performance. For each of the levels, constructed in such a way, the minimum, maximum and average values of each model parameter have been determined. The new boundaries of the model parameters are constructed in a way that the new minimum is lower but close to the minimum of the top level, and the new maximum is higher but close to the maximum of the top level. Table 1 presents previously used broad boundaries for each model parameter according to [6] as well as new boundaries proposed based on PMPG when applying MpGA. Additionally, Table 1 consists of intuitionistic fuzzy estimations, obtained based on (2).

Table 1. Model parameters boundaries for MpGA

MpGA			μ_{2s}	μ_{2x}	k_s	k_E	Y_{sx}	Y_{Ex}	k^La	Y_{Os}	Y_{Ox}
GGAP = 0.9	previously used	LB	0.9	0.05	0.08	0.5	0.3	1	0.001	0.001	0.001
		UB	1	0.15	0.15	0.8	10	10	300	1000	1000
	advisable after procedure application	LB	0.90	0.11	0.14	0.79	0.40	1.4	60	490	10
		UB	0.93	0.15	0.15	0.80	0.42	2	120	920	910
	degrees of validity of p	$M_1(p)$	0.90	0.73	0.93	0.99	0.13	0.14	0.20	0.49	0.01
	degree of non-validity of p	$N_1(p)$	0.07	0.00	0.00	0.00	0.86	0.80	0.60	0.08	0.09
GGAP = 0.5	previously used	LB	0.9	0.05	0.08	0.5	0.3	1	0.001	0.001	0.001
		UB	1	0.15	0.15	0.8	10	10	300	1000	1000
	advisable after procedure application	LB	0.90	0.11	0.13	0.79	0.39	1.5	60	470	220
		UB	0.94	0.15	0.15	0.80	0.42	2	120	930	810
	degrees of validity of p	$M_2(p)$	0.90	0.73	0.87	0.99	0.13	0.15	0.20	0.47	0.22
	degree of non-validity of p	$N_2(p)$	0.06	0.13	0.00	0.00	0.86	0.80	0.60	0.07	0.19
GGAP = 0.1	previously used	LB	0.9	0.05	0.08	0.5	0.3	1	0.001	0.001	0.001
		UB	1	0.15	0.15	0.8	10	10	300	1000	1000
	advisable after procedure application	LB	0.91	0.10	0.12	0.75	0.39	1.4	60	510	220
		UB	0.97	0.14	0.15	0.80	0.43	1.9	120	910	640
	degrees of validity of p	$M_3(p)$	0.91	0.67	0.80	0.94	0.13	0.13	0.20	0.50	0.21
	degree of non-validity of p	$N_3(p)$	0.03	0.00	0.00	0.00	0.86	0.80	0.60	0.09	0.36

Table 2 presents the boundaries (low LB and up UB) for the *strong optimistic*, *optimistic*, *average*, *pessimistic* and *strong pessimistic* prognoses for the performances of MpGA algorithm, obtained based on intuitionistic fuzzy estimations (2) and formula (3)-(7).

Investigated MpGA has been again applied for parameter identification of *S. cerevisiae* fed-batch cultivation involving newly proposed according to Table 1 boundaries at GGAP = 0.9, GGAP = 0.5 and GGAP = 0.1. Several runs have

Table 2. Prognoses for MpGA performance

	μ_S		μ_X		k_S		k_E		Y_{SX}		Y_{EX}		k^a		Y_{OS}		Y_{OE}	
	LB	UB	LB	UB	LB	UB	LB	UB	LB	UB	LB	UB	LB	UB	LB	UB	LB	UB
$V_{strong\ opt}$	1.00	1.00	0.15	0.15	0.15	0.15	0.80	0.80	1.03	1.10	3.64	4.88	146.40	235.20	864.85	999.50	397.68	993.84
V_{opt}	0.91	0.97	0.11	0.15	0.14	0.15	0.79	0.80	0.40	0.43	1.50	2.00	60.00	120.00	500.00	930.00	220	910.00
V_{aver}	0.90	0.95	0.11	0.14	0.13	0.15	0.78	0.80	0.39	0.42	1.40	2.00	60.00	120.00	486.67	920.00	150	786.67
V_{pes}	0.90	0.93	0.10	0.13	0.12	0.15	0.75	0.80	0.39	0.42	1.30	2.00	60.00	120.00	470.00	910.00	10	640.00
$V_{strong\ pes}$	0.74	0.85	0.05	0.13	0.10	0.15	0.73	0.80	0.01	0.01	0.03	0.08	2.40	19.20	115.15	778.60	0.48	471.74

been performed in order reliable results to be obtained. Table 3 presents the average values of the objective function, convergence time and model parameters when MpGA has been executed at three investigated here values of GGAP before and after the application of PMPG.

Table 3. Results from model parameter identification before and after PMPG

Parameter	GGAP = 0.9		GGAP = 0.5		GGAP = 0.1	
	before PMPG	after PMPG	before PMPG	after PMPG	before PMPG	after PMPG
J	0.0221	0.221	0.0221	0.0221	0.0222	0.0221
CPU time, s	159.67	148.91	98.96	90.51	34.16	32.13
μ_S, 1/h	0.91	0.90	0.91	0.91	0.94	0.91
μ_X, 1/h	0.12	0.13	0.13	0.13	0.11	0.13
k_S, g/l	0.15	0.15	0.15	0.15	0.14	0.15
k_E, g/l	0.80	0.80	0.80	0.80	0.79	0.80
Y_{SX}, g/g	0.41	0.41	0.41	0.41	0.42	0.41
Y_{EX}, g/g	1.64	1.77	1.69	1.76	1.45	1.70
k^a, 1/h	84.24	93.74	94.40	89.37	95.57	88.86
Y_{OS}, g/g	669.65	742.89	742.47	708.43	743.15	708.15
Y_{OE}, g/g	334.89	404.11	516.92	468.28	458.77	475.75

The applied procedure for model parameter genesis reduces the convergence time of MpGA with 6 to almost 10% but saving the model accuracy. Moreover, the results hit the top level of presentation and have one and same reduced objective function, thus showing good effectiveness of proposed procedure for purposeful model parameter genesis when MpGA is applied.

Table 4 lists the number and type of the estimations assigned to each of the parameters for three values of GGAP when MpGA is applied and before and after the PMPG.

Table 4. Model parameter estimations before and after PMPG

	GGAP = 0.9		GGAP = 0.5		GGAP = 0.1	
	before PMPG	after PMPG	before PMPG	after PMPG	before PMPG	after PMPG
strong opt	3	4	4	4	2	4
opt	6	4	5	5	6	5
aver	0	1	0	0	1	0
pes	0	0	0	0	0	0
strong pes	0	0	0	0	0	0

As seen form Table 4, there are no any *strong pessimistic* and *pessimistic* prognoses. In four of the cases there are 4 *strong optimistic* prognoses, and in three of them the next 5 prognoses are *optimistic* these are the cases of GGAP = 0.5 before and after PMPG and GGAP = 0.1 after PMPG. In these three distinguished as the most reliable cases, the value of the objective function is equal to the lowest one that means they are with the highest achieved degree of accuracy. But if one compares the time, the MpGA with GGAP = 0.1 after PMPG is about three times faster than MpGA with GGAP = 0.5 before and after PMPG and about 5 times faster than the slowest case of GGAP = 0.9 before PMPG. Thus, based on the intuitionistic fuzzy estimations of the model parameters and further constructed prognoses, MpGA with GGAP = 0.1 and after the procedure for the purposeful model parameter genesis is distinguished as more reliable algorithm if one would like to obtained results with a high level of relevance and for less computational time.

Fig. 1 shows results from experimental data and model prediction, respectively, for biomass (a), ethanol (b), substrate (c) and dissolved oxygen (d) when the procedure for the purposeful model parameter genesis has been applied for MpGA with GGAP = 0.1.

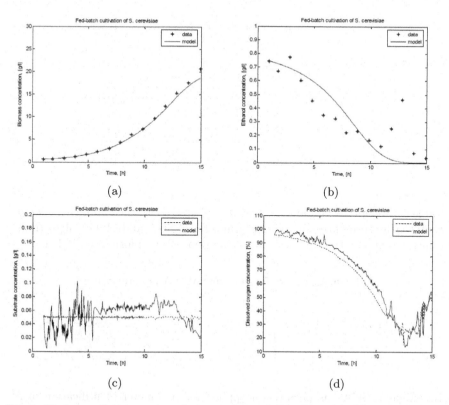

Fig. 1. Model prediction compared to experimental data when MpGA at GGAP = 0.1 has been applied

The obtained results show that the highest achieved model accuracy can be reached using MpGA with GGAP = 0.1 for much less computational time additionally reduced after the application of purposeful model parameter genesis procedure.

4   Assess the Performance of Standard SGA towards MpGA at GGAP= 0.1

In this section, distinguished as the most reliable and the fastest one MpGA at GGAP = 0.1 is going to be compared to SGA at GGAP = 0.1. Before applying AAQP procedure towards SGA and MpGA at GGAP = 0.1, plenty of runs of SGA in broad range has been performed. Based on the data collected, a new, narrow range for the parameters variation have been defined in a way described for MpGA. Two intervals of model parameters variation with corresponding low boundaries (LB) and up boundaries (UB), as well as the degrees of validity/non-validity for both algorithms accorgind (2) are shown in Table 5.

Table 5. Model parameters boundaries for SGA and MpGA at GGAP = 0.1

			μ_{2S}	μ_{2E}	k_S	k_E	Y_{SX}	Y_{EX}	k^a	Y_{OS}	Y_{OE}
SGA at GGAP = 0.1	previously used	LB	0.9	0.05	0.08	0.5	0.1	0.1	0.001	0.001	0.001
		UB	1	0.15	0.15	0.8	3	10	300	1000	1000
	advisable after procedure application	LB	0.9	0.07	0.1	0.7	0.38	1	30	240	35
		UB	1	0.15	0.15	0.8	0.45	2.1	130	980	240
	degrees of validity of p	$M_1(p)$	0.90	0.47	0.67	0.88	0.13	0.10	0.10	0.24	0.04
	degree of non-validity of p	$N_1(p)$	0.00	0.00	0.00	0.00	0.85	0.79	0.57	0.02	0.76
MpGA at GGAP = 0.1	previously used	LB	0.9	0.05	0.08	0.5	0.1	0.1	0.1	0.001	0.001
		UB	1	0.15	0.15	0.8	3	10	300	1000	1000
	advisable after procedure application	LB	0.91	0.1	0.12	0.75	0.39	1.4	60	510	220
		UB	0.97	0.14	0.15	0.8	0.43	1.9	120	910	640
	degrees of validity of p	$M_2(p)$	0.91	0.67	0.80	0.94	0.13	0.14	0.20	0.51	0.22
	degree of non-validity of p	$N_2(p)$	0.03	0.07	0.00	0.00	0.86	0.81	0.60	0.09	0.36

Table 6 presents the boundaries (low LB and up UB) for the *optimistic, average* and *pessimistic* prognoses for the performances of SGA and MpGA algorithms, obtained based on intuitionistic fuzzy estimations (2) and formula (6)-(10).

Table 6. Prognoses for SGA and MpGA performance

	μ_{2S}		μ_{2E}		k_S		k_E		Y_{SX}		Y_{EX}		k^a		Y_{OS}		Y_{OE}	
	LB	UB	LB	UB	LB	UB	LB	UB	LB	UB	LB	UB	LB	UB	LB	UB	LB	UB
V_{opt}	0.91	1.00	0.10	0.15	0.12	0.15	0.75	0.80	0.39	0.45	1.40	2.10	60.00	130.00	510.00	980.00	220.00	640.00
V_{aver}	0.91	0.99	0.09	0.15	0.11	0.15	0.73	0.80	0.39	0.44	1.20	2.00	45.00	125.00	375.00	945.00	127.50	440.00
V_{pes}	0.90	0.97	0.07	0.14	0.10	0.15	0.70	0.80	0.38	0.43	1.00	1.90	30.00	120.00	240.00	910.00	35.00	240.00

Investigated SGA has been again applied for parameter identification of *S. cerevisiae* fed-batch cultivation involving newly proposed according to Table 5 boundaries at GGAP = 0.1. Several runs have been performed in order reliable results to be obtained. Table 7 presents the average values of the objective function, convergence time and model parameters when SGA and MpGA have been executed at GGAP = 0.1 before and after the application of PMPG. Values for the MpGA at GGAP = 0.1 are equal to those presented in Table 3, but are also listed here for a completeness.

Table 7. Results from model parameter identification before and after PMPG

Parameter	SGA at GGAP = 0.1		MpGA at GGAP = 0.1	
	before PMPG	after PMPG	before PMPG	after PMPG
J	0.0225	0.0222	0.0222	0.0221
CPU time, s	25.98	18.45	34.16	32.13
μ_{2S}, 1/h	0.96	0.95	0.94	0.91
μ_{2E}, 1/h	0.11	0.12	0.11	0.13
k_S, g/l	0.13	0.12	0.14	0.15
k_E, g/l	0.75	0.79	0.79	0.8
Y_{SX}, g/g	0.41	0.4	0.42	0.41
Y_{EX}, g/g	1.48	1.67	1.45	1.7
k^a_a, 1/h	79.31	94.03	95.57	88.86
Y_{OS}, g/g	627.78	729.57	743.15	708.15
Y_{OE}, g/g	187.09	192.43	458.77	475.75

It is worth to note that in both considered here kinds of GA, running of algorithms in narrow range leads to expecting decrease of the convergence time while even improving the model accuracy. Running SGA at GGAP = 0.1 in narrow range reduces the computation time 1.41 times compared to SGA in broad range. Running MpGA at GGAP = 0.1 in narrow range reduces the computation time 1.06 times compared to MpGA in broad range. But it is obvious that the fastest one SGA at GGAP = 0.1 in narrow range is almost twice faster (exactly 1.85 times) compared to the slowest one MpGA at GGAP = 0.1 in broad range.

Table 8 lists the number and type of the estimations assigned to each of the parameters for SGA and MpGA at GGAP = 0.1 before and after the PMPG.

Table 8. Model parameter estimations before and after PMPG

	SGA at GGAP = 0.1		MpGA at GGAP = 0.1	
	before PMPG	after PMPG	before PMPG	after PMPG
opt	8	8	9	9
aver	1	1	0	0
pes	0	0	0	0

As seen form Table 8, there are no any *pessimistic* prognoses. It is worth to note that MpGA at GGAP = 0.1 in both ranges is assessed with only *optimistic* prognoses. SGA at GGAP = 0.1 in both ranges is very close to MpGA performance with 8 *optimistic* and only 1 *average* prognoses. Among the two leaders, MpGA at GGAP = 0.1 in narrow range is 1.06 times faster compared to MpGA in broad range. Moreover, the results when MpGA at GGAP = 0.1 in narrow range is applied hit the highest level of model accuracy, thus showing good effectiveness of proposed procedure for purposeful model parameter genesis. Thus, based on the intuitionistic fuzzy estimations of the model parameters and further constructed prognoses, MpGA with GGAP = 0.1 and after the procedure for the purposeful model parameter genesis is distinguished as the most reliable algorithm if one would like to obtained results with a high level of relevance.

Due to the fact that again MpGA with GGAP = 0.1 in narrow range is distinguished as the leader, results repeated those presented in Fig. 1 from experimental data and model prediction, respectively, for biomass, ethanol, substrate and dissolved oxygen.

5   Conclusions

In this investigation intuitionistic fuzzy logic has been implemented in order to assess the quality of genetic algorithms performance for the purposes of parameter identification of *S. cerevisiae* fed-batch cultivation. Aiming to save obtained promising results, namely less convergence time at kept and even improved model accuracy, intuitionistic fuzzy logic overbuilds the results obtained after the application of recently developed procedure for purposeful model parameter genesis. This procedure has been here applied to MpGA at three different values of GGAP as the most sensitive genetic algorithms parameter. After the implementation of intuitionistic fuzzy logic for obtaining of intuitionistic fuzzy estimations of model parameters and further for construction of *strong optimistic, optimistic, average, pessimistic* and *strong pessimistic* prognoses for the algorithm performances, results have been compared and MpGA with GGAP = 0.1 after the procedure for purposeful model parameter genesis application has been distinguished as more reliable. Among the distinguished three leaders, MpGA with GGAP = 0.1 after PMPG is more than three times faster than MpGA with GGAP = 0.5 before and after PMPG saving the highest achieved values of model accuracy.

Further distinguished as the most reliable and the fastest one MpGA at GGAP = 0.1 has been compared to SGA at GGAP = 0.1. AAQP has been applied to assess the performance of both standard algorithms. As a result MpGA at GGAP = 0.1 after PMPG has been distinguished as the most reliable algorithm for parameter identification of *S. cerevisiae* fed-batch cultivation with only *optimistic* prognoses and the highest model accuracy achieved.

Presented here cross-evaluation based on IFL and applied both for assessment of the influence of GGAP to the algorithm performance, as well as for the comparison of two algorithms, demonstrates the workability of intuitionistic fuzzy estimations to assist in assessment of quality of GA performance. Thus, the estimations based on intuitionistic fuzzy logic might be considered as an appropriate tool for reliable assessment for other genetic algorithm parameters, for different optimization algorithms as well as to be applied to various objects of parameter identification.

Acknowledgements. This work is partially supported by National Science Fund of Bulgaria, grants DID 02-29 and DMU 03-38.

References

1. Goldberg, D.: Genetic algorithms in search, optimization and machine learning. Addison-Wiley Publishing Company, Massachusetts (1989)
2. Pencheva, T., Roeva, O., Hristozov, I.: Functional state approach to fermentation processes modelling. In: Tzonkov, Hitzmann, B. (eds.) Prof. Marin Drinov. Academic Publishing House, Sofia (2006)
3. Jones, K.: Comparison of genetic algorithms and particle swarm optimization for fermentation feed profile determination. In: Proceedings of the CompSysTech 2006, Veliko Tarnovo, Bulgaria, June 15-16, pp. IIIB.8-1–IIIB.8-7 (2006)
4. Adeyemo, J., Enitian, A.: Optimization of fermentation processes using evolutionary algorithms - a review. Scientific Research and Essays 6(7), 1464–1472 (2011)
5. Angelova, M., Tzonkov, S., Pencheva, T.: Genetic algorithms based parameter identification of yeast fed-batch cultivation. In: Dimov, I., Dimova, S., Kolkovska, N. (eds.) NMA 2010. LNCS, vol. 6046, pp. 224–231. Springer, Heidelberg (2011)
6. Schuegerl, K., Bellgardt, K.-H. (eds.): Bioreaction engineering, modeling and control. Springer, Heidelberg (2000)
7. Gupta, D., Ghafir, S.: An overview of methods maintaining diversity in genetic algorithms. International Journal of Emerging Technology and Advanced Engineering 2(5), 56–60 (2012)
8. Angelova, M., Pencheva, T.: Improvement of multi-population genetic algorithms convergence time. Monte Carlo Methods and Application, 1–10 (2013)
9. Angelova, M., Atanassov, K., Pencheva, T.: Purposeful model parameters genesis in simple genetic algorithms. Computers and Mathematics with Applications 64, 221–228 (2012)
10. Atanassov, K.: Intuitionistic fuzzy sets. Springer, Heidelberg (1999)
11. Atanassov, K.: On intuitionistic fuzzy sets theory. Springer, Berlin (2012)
12. Pencheva, T., Angelova, M., Atanassov, K.: Intuitionistic fuzzy logic implementation to assess genetic algorithms quality. Submitted to Biochemical Engineering Journal
13. Chipperfield, A.J., Fleming, P., Pohlheim, H., Fonseca, C.M.: Genetic algorithm toolbox for use with MATLAB, Users guide, version 1.2. Dept. of Automatic Control and System Engineering, University of Sheffield, UK (1994)

References

1. Goldberg, D.: Genetic algorithms in search, optimization and machine learning. Addison-Wesley Publishing Company, Massachusetts (1989)

A Graph Optimization Approach to Item-Based Collaborative Filtering

Borzou Rostami, Paolo Cremonesi, and Federico Malucelli

Politecnico di Milano - DEI, Pizza Leonardo da Vinci, 32 Milan, Italy
{rostami,malucell}@elet.polimi.it, paolo.cremonesi@polimi.it

Abstract. Recommender systems play an increasingly important role in online applications characterized by a very large amount of data and help users to find what they need or prefer. Various approaches for recommender systems have been developed that utilize either demographic, content, or historical information. Among these methods, item-based collaborative filtering is one of most widely used and successful neighborhood-based collaborative recommendation approaches that compute recommendation for users using the similarity between different items. However, despite their success, they suffer from the lack of available ratings which leads to poor recommendations. In this paper we apply a bi-criterion bath optimization approach on a graph representing the items and their similarity. This approach introduces additional similarity links by combining two or more existing links and improve the similarity matrix between items. The two criteria take into account on the one hand the distance between items on a the graph (min sum criterion), on the other hand the estimate of the information reliability (max min criterion). Experimental results on both explicit and implicit datasets shows that our approach is able to burst the accuracy of existing item-based algorithms and to outperform other algorithms.

Keywords: Item-based collaborative filtering, Similarity measure, Bi-criterion path problem, MinSum-MaxMin optimization.

1 Introduction

Recommender Systems (RS) play an increasingly important role in online applications characterized by a very large amount of data - e.g., multimedia catalogs of music, products, news, images, or movies. Their goal is to filter a large amount of information to identify the items that are likely to be more interesting and attractive to a user. Recommendations are inferred on the basis of different user profile characteristics, including either explicit or implicit ratings on a sample of suggested elements. Explicit ratings confidently represent the user opinion; For instance, in a movie recommendation application users can enter ratings explicitly after watching a movie, giving their opinion on this movie. On the other hand, implicit ratings are inferred by the system on the basis of the user-system interaction, which might not match the user opinion. For instance, the system is able to monitor whether a user has watched a live program on a certain channel or whether the user has uninterruptedly watched a movie. Despite explicit ratings are more reliable than implicit ratings in representing the actual user interest towards an item, their collection can be annoying from the user's perspective. The

S. Fidanova (Ed.): *Recent Advances in Computational Optimization*, SCI 470, pp. 15–30.
DOI: 10.1007/978-3-319-00410-5_2   © Springer International Publishing Switzerland 2013

problem of recommending items has been studied extensively, and two main paradigms have emerged. Content-based recommendation systems try to recommend items similar to those a given user has liked in the past [4,5], whereas systems designed according to the collaborative recommendation paradigm identify users whose preferences are similar to those of the given user and recommend items they have liked [14]. There are two major approaches to collaborative filtering: (i) neighborhood models and (ii) dimensionality reduction models.

Neighborhood models base their prediction on the similarity relationships between either users or item. Algorithms based on the similarity between users predict a user's preference on an item based on the ratings that item has received from similar users. On the other hand, algorithms based on the similarity between items compute the user's preference for an item based on his/her own ratings on similar items. The latter is usually the preferred approach, as it usually performs better in terms of accuracy, while also being more scalable [23].

Item-based systems suffer from the lack of available ratings. When the rating data is sparse, it is possible to have items with few ratings in common; therefore, similarity weights may be computed using only a small number of ratings and consequently the item-based approach will make predictions using a very limited number of neighbors, resulting in a biased recommendation.

Dimensionality reduction is one of the common approaches used to overcome the problems of sparsity and scalability in CF. Decomposition of a user-rating matrix [8,11,26] and decomposition of a sparse similarity matrix [10] are essentially two ways in which dimensionality reduction can be used to improve recommender systems.

Graph-based approaches also have been introduced to overcome the problems arising in neighborhood collaborative filtering due to sparsity. These approaches make use of a graph where nodes correspond to users, items or both, and edges represent the interactions or similarities between users and items. Recommendations are then induced by "transitive associations", that is suitable paths in the graph that have the role to reduce graph sparsity. The transitive associations can be used to recommend items in two different ways. In a first approach, the proximity of a user u to an item i in the graph is used directly to evaluate the rating of u for i [16,28]. Following this idea, the items recommended to u by the system are those that are the closest to u in the graph. The second approach considers the proximity of two item nodes in the graph as a measure of similarity, and uses this similarity as the weights of a neighborhood-based recommendation method [9,19].

In path-based similarity, the distance between two nodes of the graph is evaluated as a function of the number of paths connecting the two nodes, as well as the length of these paths. A recommendation approach that computes the similarity between two users based on their shortest distance in a graph is the one described in [1]. In this method, the data is modeled as a directed graph whose nodes are users, and in which edges are determined based on the notions of horting and predictability. The number of paths between a user and an item in a bipartite graph can also be used to evaluate their compatibility [16]. This method of computing distances between nodes in a graph is known as the Katz measure [17]. Another direction in collaborative filtering research combines user-based and item-based approaches. For example, [29] clusters the user data and applies intra-cluster smoothing to reduce sparsity. Also in [12] a procedure for computing similarities between elements of a database has been presented which is based on a Markov-chain model of random walk through a graph representation of the database. The presented similarity measures can be used in order to compare items belonging to database tables that are not necessarily directly connected.

Based on the above discussion and in order to overcome sparsity in CF, we present an optimization approach in item-based CF which is based on the item graph [27]. We define a weighted graph where nodes correspond to items and arcs are similarity link between items. For each arc a real numbers is assigned representing the reliability of the arc. In order to find a new similarity link between two items with unknown similarity in the item graph, first we formulate the problem as a bi-criterion path optimization problem [24]. By applying an efficient polynomial algorithm [13] for bi-criterion path optimization we find a subset of "efficient" paths in the graph as a best candidate set of paths between these two nodes. Eventually we use the best path in the candidate set to assign the similarity weight to the new link. The general framework of our work, like other graph-based models, is based on finding one or more paths between two items. However, in our method the similarity between items is found by considering not only the distance between two items or the length of the path joining them but also by taking to account the "reliability" of the path which connects them.

The rest of the paper is organized as follows: Section 2 describes some collaborative algorithms considered in our study, to provide the needed technical background for the following sections. In section 3, we first formulate our problem as an optimization problem then we present an efficient algorithm to find the bi-criterion path problem in networks and apply this algorithm to our problem. The experimental results will be provided in section 4. In section 5 the conclusion and discussion will be presented.

2   Collaborative Filtering

Collaborative filtering recommends items on the basis of the ratings provided by groups of users. The main input to collaborative algorithms is the *user rating matrix*, where each element r_{ui} is user u's rating on item i (e.g., $r_{ui} \in \{1, ..., 5\}$ or $r_{ui} \in \{like, dislike\}$). There are two major approaches to collaborative filtering: (i) neighborhood models and (ii) dimensionality reduction models.

2.1   Neighborhood Models

Neighborhood models base their prediction on the similarity relationships between either users or items. Algorithms based on the similarity between users predict a user's preference on an item based on the ratings that item has received from similar users. On the other hand, algorithms based on the similarity between items compute the user's preference for an item based on his/her own ratings on similar items. The latter is usually the preferred approach (e.g., [23]), as it usually performs better in terms of accuracy, while also being more scalable. Both of these advantages are due to the fact that the number of items is typically smaller than the number of users. Another advantage of the latter algorithms is that the reason why a specific recommendation was made to a user can be explained in terms of the items previously rated by him/her. In addition, basing the model on items (rather than on users) allows a seamless handling of users and ratings that are new to the model.

Item based recommendation algorithm contains two main phases, the model building phase and the prediction phase. In the model building phase, the similarities between each pair of items are computed and for each particular item i, the algorithm will store its k most similar items and their similarity values with i. Therefore the similarity weights play a double role in the neighborhood-based recommendation methods: 1) they allow for the selection of trusted neighbors whose ratings are used in the prediction, and 2) they provide the means to give more or less importance to these neighbors in the prediction.

The basic idea in similarity computation between two items i and j is to first consider the users who have rated both of the items and then apply a similarity technique to determine the similarity weight. The similarity s_{ij} between item i and item j is measured as the tendency of users to rate items i and j similarly and can be defined in many different ways. Here are three common approaches in [23].

– Cosine

$$s_{ij} = \frac{\sum\limits_{u} r_{ui} r_{uj}}{\sqrt{\sum\limits_{u} r_{ui}^2} \sqrt{\sum\limits_{u} r_{uj}^2}} \tag{1}$$

– Adjusted cosine

$$s_{ij} = \frac{\sum\limits_{u} (r_{ui} - \bar{r}_u)(r_{uj} - \bar{r}_u)}{\sqrt{\sum\limits_{u} (r_{ui} - \bar{r}_u)^2} \sqrt{\sum\limits_{u} (r_{uj} - \bar{r}_u)^2}} \tag{2}$$

– Pearson correlation

$$s_{ij} = \frac{\sum\limits_{u} (r_{ui} - \bar{r}_i)(r_{uj} - \bar{r}_j)}{\sqrt{\sum\limits_{u} (r_{ui} - \bar{r}_i)^2} \sqrt{\sum\limits_{u} (r_{uj} - \bar{r}_j)^2}} \tag{3}$$

where $\bar{r}_u$ is the average rating of the u-th user and $\bar{r}_i$ (respectively, $\bar{r}_j$) is the average rating of the i-th (respectively, j-th) item. Summations in the cosine similarity are computed over all the users. On the contrary, summations in both the adjusted cosine and the Pearson similarities are computed only on users who have rated both items i and j – the common raters – and the similarity is set to zero for pairs of items with no common raters. In the typical case of a very sparse dataset, it is likely that some pairs of items will have a poor support (i.e., a small number of common raters), leading to a non-reliable similarity measure. This is why, if n_{ij} denotes the number of common raters and s_{ij} the similarity between item i and item j, we can define the shrunk similarity d_{ij} as the coefficient

$$d_{ij} = \frac{n_{ij}}{n_{ij} + \lambda} s_{ij} \tag{4}$$

where λ is a shrinking factor. A good value for λ is 100 [18].

Neighborhood models are further enhanced by means of a kNN (k-nearest-neighborhood) approach: when predicting a rating $\hat{r}_{ui}$ for user u on item i, only the k items rated by u that are the most similar to i are considered. The kNN approach discards the items that are poorly correlated to the target item, thus decreasing noise for improving the quality of recommendations. We denote the set of k items rated by user u, and most similar to i, as $\mathcal{D}^k(u;i)$. We have focused our attention on two item-based neighborhood algorithms, i.e., Non-Normalized Cosine Neighborhood and Direct Relations.

2.1.1   Non-Normalized Cosine Neighborhood (NNCosNgbr)

The NNCosNgbr algorithm predicts the rating $\hat{r}_{ui}$ for user u on item i as the weighted average of the ratings of similar items. Before computing the weighted average, we normalize the ratings by removing different biases which mask the more fundamental relations between items. The bias associated with the rating of user u to item i is denoted by b_{ui} and it is subtracted from rating r_{ui}. Such biases include item-effects, which represent the fact that certain items tend to receive higher ratings than others, and user-effects, which represent the tendency of certain users to rate higher than others. For instance, a simple formulation for the bias could be

$$b_{ui} = \bar{r} + (\bar{r}_u - \bar{r}) + (\bar{r}_i - \bar{r}) \tag{5}$$

where $\bar{r}$ is the average of all the ratings in the user rating matrix, $\bar{r}_u$ is the average rating of the u-th user and $\bar{r}_i$ is the average rating of the i-th item. By removing the bias effects, the rating estimation is

$$\hat{r}_{ui} = b_{ui} + \frac{\sum_{j \in \mathcal{D}^k(u;i)} d_{ij} (r_{uj} - b_{uj})}{\sum_{j \in \mathcal{D}^k(u;i)} d_{ij}} \tag{6}$$

where d_{ij} is computed as (4), and s_{ij} is measured as the Pearson correlation coefficient.

Notice that the denominator in (6) forces the predicted rating values to fall within a defined range, e.g., $[1 \ldots 5]$ for a typical star-rating system. However, for a top-N recommendation task, exact rating values are not necessary. We simply want to rank items by their appeal to the user. In such a case, we can simplify the formula by removing the denominator. A consequential benefit of this is that items with many similar neighbors, that is with a high value of $\sum_{j \in \mathcal{D}^k(u;i)} d_{ij}$, which means in turn that we have a high confidence in the recommendation, have higher rankings. Therefore, we propose to rank items with the following rating estimation

$$\hat{r}_{ui} = b_{ui} + \sum_{j \in \mathcal{D}^k(u;i)} d_{ij} (r_{uj} - b_{uj}) \tag{7}$$

Here $\hat{r}_{ui}$ does not represent a proper rating, but is rather a value we can use to rank the items according to user u's taste. We should note that similar non-normalized neighborhood rules have been mentioned by others [18,8].

2.1.2 Direct Relations (DR)

An alternative and simple way of computing the similarity between pair of items i and j in (7) is to count the number of users that rated both items, without any normalization factor

$$d_{ij} = \# \; users \; rating \; both \; items \qquad (8)$$

According to [2] this metric emphasizes the similarity between popular items.

2.2 Dimensionality Reduction Models

Recently, several recommender algorithms based on dimensionality reduction have been proposed. Some of the most successful realizations of dimensionality reduction models are based on *matrix factorization*. In its basic form, matrix factorization characterizes items and users by vectors of factors inferred from item rating patterns. High correspondence between item and user factors leads to a recommendation. These methods have become popular since they combine predictive accuracy with good scalability.

In dimensionality reduction models, each item i is associated with a vector $\mathbf{q}_i \in R^f$, and each user u is associated with a vector $\mathbf{p}_u \in R^f$, where f is the number of latent factors. For a given item i, the elements of $\mathbf{q}_i$ measure the extent to which the item possesses those factors, either positively or negatively. For a given user u, the elements of $\mathbf{p}_u$ measure the extent to which the user is interested in items that have high values for the corresponding factors, again, either positively or negatively. The resulting dot product, $\mathbf{q}_i^T \mathbf{p}_u$, captures the interaction between user u and item i, i.e., the user's overall interest in the item's characteristics. This approximates user u's rating of item i, leading to the estimate $r_{ui} = \mathbf{q}_i^T \mathbf{p}_u$. The major challenge is to compute the mapping of each item and user to factor vectors $\mathbf{q}_i, \mathbf{p}_u$. After the recommender system completes this mapping, it can easily estimate the rating a user will give to any item.

Dimensionality reduction models based on *matrix factorization* informally known as Singular Value Decomposition (SVD) models. Since conventional SVD is undefined in the presence of missing values, which translate to unknown user ratings, several alternative solutions have been proposed. Earlier works fill the missing ratings with baseline estimations (e.g., average user/item rating [22]). This however leads to a very large and dense user rating matrix, whose factorization might be computationally infeasible. More recent works learn the values from the known ratings through a suitable objective function which minimizes the prediction error (e.g., RMSE). The proposed objective functions are usually regularized in order to avoid over-fitting [21]. Typically, gradient descent is applied to minimize the objective function. In this work we have considered PureSVD techniques treats missing ratings as zeros and performs a traditional SVD.

2.2.1 PureSVD

PureSVD is a recently proposed latent factor algorithm [6]. Its rating estimation rule is based on conventional SVD, where unknown ratings are treated as zeros. In terms of predictive power, choosing zero is not very important, and we have received similar results with higher values. What is important is that the conventional SVD decomposition of the user rating matrix becomes feasible, since all matrix entries are now non-missing and it can be performed using highly-optimized tools for conventional SVD on sparse matrices. The user rating matrix $\mathbf{R}$ is estimated by the factorization [2]

$$\hat{R} = U \cdot \Sigma \cdot Q^T \tag{9}$$

where $\mathbf{U} \in \mathfrak{R}^{n \times f}$ and $\mathbf{Q} \in \mathfrak{R}^{m \times f}$ are two orthonormal matrices representing, respectively, the left and right singular vectors associated to the top-f singular values of $\mathbf{R}$ with the highest magnitude. The top-f singular values are stored in the diagonal matrix $\Sigma \in \mathfrak{R}^{f \times f}$. As detailed in [6], once the user rating matrix has been decomposed, the prediction rule for PureSVD can be written as

$$\hat{r}_{ui} = r_u \cdot Q \cdot q_i{}^T \tag{10}$$

where $\mathbf{r}_u$ denotes user u's vector of ratings (where unknown ratings are filled with zeros), and $\mathbf{q}_i$ represents the i-th column of $\mathbf{Q}$. Note that, similarly to (6), $\hat{r}_{ui}$ is not a proper normalized rating, but can be used to rank items according to user u's interests.

3   Optimization Model

In this section we first describe the items and relationship between them by means of a weighted graph, then we formulate the problem of finding the similarity between items with unknown relationship as a bi-criterion path problem. Suppose that $M = \{1, 2, ..., m\}$ is the set of users, $V = \{1, 2, ..., n\}$ is the set items and $R \in \mathfrak{R}^{m \times n}$ is the user-rating matrix where each entry a_{hk} gives the rating that user h gave to item$k$, if any. Moreover, suppose that we are given the item similarity matrix $S \in \mathfrak{R}^{n \times n}$. For instance, S can be obtained by Cosine similarity method. Based on similarity matrix we define a weighted graph G with vertex set V and arc set $E = \{(i, j) \in V \times V : s_{ij} > 0\}$. Each arc $(i, j) \in E$ is weighted by the similarity of two items.

Consider two non adjacent nodes a and b in the graph G that is two items with unknown similarity. Our objective is to look for a similarity weight between items a and b so as to improve the quality of the recommender system. This objective can be translated as introducing one arc between nodes a and b in the graph G. A natural way to find a new connection between two nodes in a graph is to find a path which connects these two nodes. Selecting the "best" path among all possible paths between these two nodes lead us to solve an optimization problem.

Let $N = \{(i, j) \in V \times V : s_{ij} = 0\}$ is the set of pair of items in G with unknown similarity, $(a, b) \in N$ and $\prod_{ab}$ be the set of all paths between a and b in G. The reliability of a path $P \in \prod_{ab}$ is defined by the lowest similarity weight in the path:

$$reliability(P) = \min_{(i,j) \in P} s_{ij} \tag{11}$$

This definition implies that finding a path with the maximum reliability between two items a and b, corresponds to finding a path whose arc of minimum weight s_{ij} is maximum among all paths $P \in \prod_{ab}$, giving rise to the following bottleneck path problem:

$$\max_{P \in \prod_{ab}} reliability(P) \tag{12}$$

One of the most critical issues with the previous problem is that the maximum reliability path might be too long in terms of arcs. Although in our formalization paths are only weighted by the value of the arc of minimum reliability, in practice it also makes sense to require that the paths should be short in terms of the number of "hops" in the path. The realization of this idea yields the following optimization problem:

$$\min_{P\in\Pi_{ab}} |P| \tag{13}$$

Where $|P|$ denotes the cardinality of P. By combining the two objective functions considered above we define the best path $P^* \in \Pi_{ab}$.

Definition 1. *A path would be selected as a "best" path if it satisfies in the two following criteria:*
Criterion 1: *Selected path must have the maximum reliability in* Π_{ab}.
Criterion 2: *Selected path must include the minimum number of "hops".*

According to criterion 1 and criterion 2 we must find minimum cardinality path with maximum reliability in G which is a kind of bi-criterion path optimization problem and called MinSum-MaxMin bi-criterion path optimization problem [13].

As it is highly unlikely to find a path from node a, to node b which achieves both the minimum cardinality and maximum reliability, we have to settle with something less, namely finding the set of efficient paths from a to b.

Definition 2. *A path $P \in \Pi_{ab}$ is efficient if and only if no other path $P' \in \Pi_{ab}$ has a better value for one criterion and not worse value for the other one.*

A path which is not efficient is thus dominated by at least one efficient path. Hence two efficient paths are equivalent if and only if their value agree for both criteria.

Definition 3. *A set $C_{ab} \subset \Pi_{ab}$ of efficient paths is complete, if any path $P' \notin C_{ab}$ is either dominated or equivalent to at least one efficient path $P \in C_{ab}$.*

Definition 4. *A complete set C_{ab} is minimal if and only if no two of its efficient paths are equivalent.*

In order to find a new similarity link consider a given graph $G = (V,E)$ with reliability s_{ij} of its arcs, initial vertex a and terminal vertex b (e.i., $(a,b) \in N$) as nodes with unknown similarity weight. Let $C_{ab}^* = \{P_1,P_2,...,P_\ell\}$ be the complete set associated with these two nodes and $\lambda_1,...,\lambda_\ell$ and $\mu_1,...,\mu_\ell$ be the reliability and cardinality of the paths $P_1,P_2,...,P_\ell$ respectively. Define

$$\Theta_{ab} = \max\{\theta_i : \theta_i = \lambda_i/\mu_i,\ i = 1,2,...,\ell\} \tag{14}$$

Then the similarity between items a and b, ξ_{ab}, can be defined as follows:

$$\xi_{ab} = \lambda_i : \theta_i = \Theta_{ab} \tag{15}$$

3.1 Proposed Algorithm

As we defined before, a path $P \in \prod_{ab}$ is efficient if and only if no other path $P' \in \prod_{ab}$ has a better value for one criterion and not worse value for the other one. In another word, a path P is efficient for the MINSUM-MAXMIN problem if and only if $\prod_{ab}$ contains no path P' for which the cardinality is smaller and the reliability of the cheapest arc is not smaller or for which the cardinality is not greater and the reliability of the cheapest arc is larger. For any $(a, b) \in N$, a minimal complete set contains at most $|E|$ efficient paths. The following algorithm, which is the revised version of the one in [13], yield the set of all the efficient paths between a and b, new similarity weight, ξ_{ab}, and exploits the structure of the MINSUM-MAXMIN problem to avoid recomputing some labels when it is not compulsory.

Labeling algorithm:
1. Start.
   - Consider S as input similarity matrix and for all $(a, b) \in N$ and $a < b$ repeat:
2. Initialization.
   - Define an ordered set T with $|T| = n$ such that $T(1) = a, T(n) = b, T(a) = 1, T(b) = n$ and $T(j) = j$ for all $j \in V, 1 < j < n$ and $j \neq a, b$.
   - Define two sets of label λ and μ. Let $\lambda_a = 0, \mu_a = \infty, \lambda_j = \infty$ and $\mu_j = 0$ for $j \in V, j \neq a$.
   - Set $p_j = 0$ for $j \in V$ and $s_{min} = -1$.
   - $\Theta'_{ab} = \emptyset$.
3. Smallest label and ending test.
   Compute $\underline{\lambda} = Min\{\lambda_j : j \in T\}$.
   - If $\underline{\lambda} = \infty$ go to 7, all the efficient path between a and b have been found.
   - Otherwise compute $\overline{\mu} = Max\{\mu_j : j \in T, \lambda_j = \underline{\lambda}\}$ and select the vertex v_k such that $k = Max\{j : j \in T, \lambda_j = \underline{\lambda}, \mu_j = \overline{\mu}\}$. If $k = b$ go to 5, otherwise $T = T - \{k\}$ and go to 4.
4. New label.
   For all $j \in T$ such that $(j, k) \in E$ and $s_{kj} > s_{min}$
   - If $\lambda_j > \lambda_k + 1$ then set $\lambda_j = \lambda_k + 1$, $\mu_j = \min\{\mu_k, s_{kj}\}$ and $p_j = k$.
   - If $\lambda_j = \lambda_k + 1$ and $\mu_j < \min\{\mu_k, s_{kj}\}$ then set $\mu_j = \min\{\mu_k, s_{kj}\}$ and $p_j = k$.
   - return to 3
5. Output of an efficient path.
   - Print λ_n and μ_n. Also set $j = p_j$, recursively print j and set $j = p_j$ until $j = 0$.
   - Set $\Theta'_{ab} = \Theta'_{ab} \cup \{\theta : \theta = \lambda_k/\mu_k\}$.
6. Suppression of arcs and updating of labels.
   - Set $s_{min} = \mu_n, V_1 = \{j : j \in V, \lambda_j < \infty, \mu_j \leq s_{min}\}$ and $T = T \cup V_1$.
   - For each $j \in V_1$ let $P_j = \{k : (k, j) \in E, k \in V - T, s_{kj} > s_{min}\}$.
     - If $P_j = \emptyset$ set $\lambda_j = \infty, \mu_j = 0$ and $p_j = 0$.
     - Otherwise set:
       - $\lambda_j = \min_{k \in P_j}\{\lambda_k + 1\}$
       - $\mu_j = \max_{k \in P_j : \lambda_j = \lambda_k + 1} \min\{\mu_k, s_{kj}\}$
       - $p_j = \max\{k : k \in P_j, \lambda_j = \lambda_k + 1, \mu_j = min(\mu_k, s_{kj})\}$
     - return to 3.

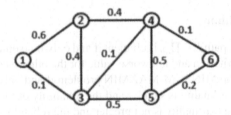

Fig. 1. Item-graph corresponding to a given similarity matrix

7. New similarity link.

$$\xi_{ab} = \lambda_t : \theta_t = \max\{\theta_i : i \in \Theta'_{ab}\}.$$

Theorem 1. *For each* $(a,b) \in N$ *the labeling algorithm yields a minimal complete set of efficient paths in* $O(|E|logn)$ *operations.*

Proof. See [13].

As an example consider the item graph in Figure 1 and suppose we are considering the introduction of a similarity link between items 1 and item 6. According to the labeling algorithm we find the minimal complete set $C^*_{16} = \{P_1, P_2\}$. Where $P_1 = (1,2,4,6)$ with $\lambda_1 = 0.1$, $\mu_1 = 3$ and $P_2 = (1,2,4,5,6)$ with $\lambda_2 = 0.2$, $\mu_1 = 4$. $\Theta_{16} = \max\{\frac{0.1}{3}, \frac{0.2}{4}\}$. By using (15) the similarity weight between items 1 and 6 is found as:

$$\xi_{16} = 0.2$$

corresponding to path P_2.

4   Experimental Results

In this section we present the quality of our graph-based optimization algorithm and the recommender algorithms presented in Section 2 on both kinds of implicit and explicit datasets. As for the explicit dataset we selected two standard datasets: *MovieLens* [20] and a subset of the *Netflix* [3]. Both are publicly available movie rating datasets which collected ratings are in a 1-to-5 star scale. We also selected an implicit dataset, Fastweb, collecting the watching behavior of TV user [2]. Table 1 summarizes their statistical properties.

Table 1. Statistical properties of Movielens, Netflix and Fastweb datasets

Dataset	Users	Items	Density
Movielens	6,040	3,883	4.26%
subset of Netflix	247,939	6,489	0.55%
Fastweb	47,526	583	0.92%

4.1 Testing Methodology

We apply two different testing methodologies for explicit and implicit datasets respectively. The testing methodology adopted in this study for explicit datasets is similar to the one described in [6] while for implicit dataset we have adopted the k-fold methodology described in [7].

4.1.1 Testing Methodology for Explicit Datasets

For each explicit dataset, known ratings are split into two subsets: training set M and test set T. The test set T contains only 5-stars ratings. So we can reasonably state that T contains items relevant to the respective users.

The original Netflix dataset was released partitioned into two parts: training-set and probe-set. Ratings in the training set M were a subset of the original Netflix training set. Some items (and the corresponding ratings) were removed because of the lack of complementary data, while other items were merged because of the different editions of the same movie existing. Ratings in the test set T were a subset of the Netflix probe-set. Again, some items and their corresponding ratings were removed or merged. Moreover, only 5-star ratings (or 1-star ratings) were retained from the Netflix probe-set, thus leading to the creation of two test sets T. The first test set only contained 5-stars and was used for the computation of recall, while the second test set only contained 1-stars and was used for the computation of fallout (see (16) and (17) for a definition of *recall* and *fallout*).

We adopted a similar procedure for the Movielens dataset [20]. We randomly subsampled 1.4% of the ratings from the dataset in order to create a probe set. The training set M contains the remaining ratings. The test set T contains all the 5-star ratings from the probe set.

The same training set was used across all the algorithms, and a standard hold-out technique was adopted for the testing methodology. For each rating in the testing set, we predicted the rating together with the ratings of an additional 1000 unrated random items. The corresponding list was sorted and recommended to the user. Thus, the testing methodology used the whole training set to build the model, and recommended a list of $1000 + 1$ items to the user.

In order to measure recall, we first trained the algorithm using the ratings in M. Then, for each item i in T that was rated 5 stars by user u, we followed these steps:

1. We randomly selected $1,000$ additional items that were not rated by user u. We assumed that the user u was not interested in most of them.
2. We predicted the ratings for the test item i and for the additional $1,000$ items.
3. We formed a top-N recommendation list by picking the N items with the largest predicted ratings.

Overall, we generated a number of recommendation lists equal to the number of elements in T. For each list we had a hit (e.g., a successful recommendation) if the test item was in the list. Therefore, the overall recall r was computed by counting the number of successful recommendations over the total number of recommendations

$$recall = \frac{\# \ times \ the \ element \ is \ in \ the \ list}{\# \ elements \ in \ T} \tag{16}$$

A similar approach was used to measure fallout, with the only difference being the composition of the test set T. In this case it only contained part of the 1-star ratings. Therefore we can reasonably state that this test set contained items that were not relevant to the users. The fallout f was defined as

$$fallout = \frac{\text{\# times the element is in the list}}{\text{\# elements in } T} \tag{17}$$

4.1.2   Testing Methodology for Implicit Dataset

In the k-fold methodology the *user rating matrix* is divided into k folds by partitioning the rows (i.e., the users) into k disjoint sets (in our tests $k = 2$): k-1 folds are used to train the algorithm while the remaining fold is used to evaluate the algorithm. The process is repeated k times, selecting a different test fold at each iteration. Within the test fold, users are tested by using a leave-one-out approach: for each tested user, each rated item is removed in turn from the user profile, and recommendations are generated on the basis of the remaining user ratings. If the removed item is recommended to the user within the first N positions we have a hit (in our tests $N \in \{1, ..., 10\}$). In implicit dataset since we only have binary ratings, is not possible to compute the *fallout*; therefore, we are forced to use classification accuracy metrics (e.g., *recall*) for the systems evaluation and find a few specific items (top-N recommendation) which are supposed to be most appealing to the user.

4.2   Results

4.2.1   Results for Explicit Datasets

To evaluate the algorithms we employed the receiver operator characteristic (ROC) curve [25] advocated for recommender system evaluation by Herlocker [15]. To create the ROC curve we vary the number of recommended items in a ranked list that we use as recommendations. ROC curves plot the miss rate (fallout) on the x-axis against the hit rate (recall) on the y-axis. Recall measures the percentage of items in the catalog interesting for the user and that the recommender system is able to suggest to the user. Fallout measures the percentage of items in the catalog not interesting for the the user and that the recommender system erroneously suggests to the user. Ideally, a good algorithm should have high recall (i.e. it should be able to recommend items of interest to the user) and low fallout (i.e. it should avoid recommending items of no interest to the user).

Figures 2 and 3 represent the results by using ROC curves. For both explicit datasets, the graph-based algorithm outperforms other algorithms, as the best results are obtained when a curve is close to the upper-left corner of the diagram (i.e., low fall-out and large recall).

The improvement of quality is more evident on the Netflix dataset. This is an expected result, as in our experiments the Netflix dataset is sparser than the Movielens dataset and traditional CF techniques suffer from the sparsity problem. For instance, the number of similarity links between items in traditional item-based algorithms – such as NNCosKnn and DR – is low if the input user rating matrix is too sparse. On the contrary, our graph-based approach is able to overcame this problem by finding additional similarity links.

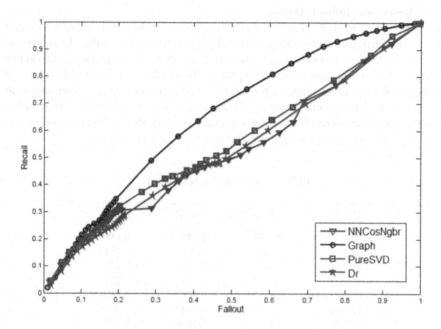

Fig. 2. Netflix data set: ROC curves

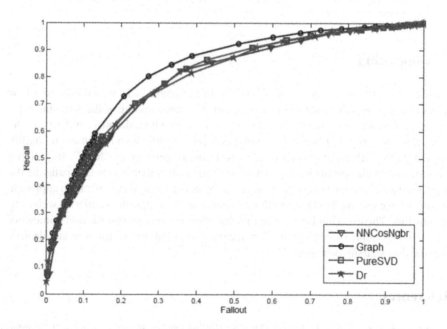

Fig. 3. MovieLens dataset: ROC curves

4.2.2 Results for Implicit Dataset

We can't employ the ROC curves to evaluate the implicit dataset, because fallout can not be computed for binary ratings. Table 2 reports the *recall* of the algorithms on the Fastweb dataset. Reported recalls show that our graph-based algorithm has better accuracy among other algorithms. For instance, the *recall* of our algorithm at $N = 10$ is about 0.53, i.e., the model has a probability of 53% to place an appealing movie in the top-10. The best algorithm among the non-graph approaches in terms of *recall* is the NNCosNgbr which reach, at $N = 10$, a recall equaling about 0.50. This means that about 50% of 5-star movies are presented in a top-10 recommendation.

Table 2. Recall for Fastweb dataset

N	Graph	NNCosNgbr	PureSVD	Dr
1	0.1934	0.1815	0.1467	0.1486
2	0.2664	0.2513	0.2067	0.2092
3	0.3214	0.3057	0.2543	0.2540
4	0.3663	0.3502	0.2942	0.2938
5	0.4037	0.3882	0.3287	0.3304
6	0.4360	0.4211	0.3593	0.3606
7	0.4628	0.4403	0.3868	0.3879
8	0.4881	0.4616	0.4112	0.4131
9	0.5093	0.4762	0.4332	0.4368
10	0.5289	0.5005	0.4538	0.4587

5   Conclusion

In order to overcome the problem of sparsity in item-based CF, we introduced a new optimization approach which is based on the graph representation of the item similarity matrix. To find a new similarity link between two items with unknown similarity, in the item graph, we proposed two optimization criteria: (i) paths with maximum reliability and (ii) paths with minimum cardinality in terms of number of "hops". By solving a bi-criterion path optimization problem we found all possible efficient paths in the graph then we chose the best similarity weight by using a simple optimization approach. Eventually we use the best path in the candidate set to assign the similarity weight to the new link. The experiments showed that our new bi-criterion optimization framework is effective in improving the prediction accuracy of collaborative filtering and dealing with the data sparsity problem.

References

1. Aggarwal, C.C., Wolf, J.L., Wu, K., Yu, P.S.: Horting hatches an egg: A new graph-theoretic approach to collaborative filtering. In: Proc. of the 5th ACM SIGKDD Int. Conf. on Knowledge Discovery and Data Mining, KDD 1999, pp. 201–212. ACM, New York (1999)

2. Bambinip, R., Cremonesi, P., Turrin, R.: A recommender system for an iptv service provider: a real large-scale production environment. In: Recommender Systems Handbook, pp. 299–331. Springer (2010)
3. Bennet, J., Lanning, S.: The netflix prize. In: Proc. of the KDD Cup and Workshop (2007)
4. Billsus, D., Pazzani, M.J.: Learning collaborative information filters. In: Proc. of the 15th Int. Conf. on Machine Learning, ICML 1998, pp. 46–54. Morgan Kaufmann Publishers Inc., San Francisco (1998)
5. Billsus, D., Pazzani, M.J.: User modeling for adaptive news access. User Modeling and User-Adapted Interaction 10(2-3), 147–180 (2000)
6. Cremonesi, P., Koren, Y., Turrin, R.: Performance of recommender algorithms on top-n recommendation tasks. In: RecSys, pp. 39–46 (2010)
7. Cremonesi, P., Turrin, R.: Analysis of Cold-Start Recommendations in IPTV Systems. In: RecSys, pp. 39–46 (2009)
8. Deshpande, M., Karypis, G.: Item-based top-N recommendation algorithms. ACM Transactions on Information Systems 22(1), 143–177 (2004)
9. Fouss, F., Renders, J.M., Pirotte, A., Saerens, M.: Random-walk computation of similarities between nodes of a graph with application to collaborative recommendation. IEEE Transactions on Knowledge and Data Engineering 19(3), 355–369 (2007)
10. Goldberg, K., Roeder, T., Gupta, D., Perkins, C.: Eigentaste: A constant time collaborative filtering algorithm. Information Retrieval 4(2), 133–151 (2001)
11. Golub, G.H., Van Loan, C.F.: Matrix computations, 3rd edn. Johns Hopkins University Press (1996)
12. Gori, M., Pucci, A.: Itemrank: A random-walk based scoring algorithm for recommender engines. In: Proc. of the 2007 IJCAI Conf., pp. 2766–2771 (2007)
13. Hansen, P.: Bicriterion path problems. In: Multiple Criteria Decision Making: Theory and Applications, pp. 109–127. Springer, Heidelberg (1980)
14. Herlocker, J., Konstan, J., Teveen, L., Riedl, J.: Evaluating collaborative filtering recommender systems. ACM Transactions on Information Systems (TOIS) 22(1), 5–53 (2004)
15. Herlocker, J.L., Konstan, J.A., Borchers, A., Riedl, J.: An algorithmic framework for performing collaborative filtering. In: Proceedings of the Conference on Research and Development in Information Retrieval (1999)
16. Huang, Z., Chen, H., Zeng, D.: Applying associative retrieval techniques to alleviate the sparsity problem in collaborative filtering. ACM Transactions on Information Systems 22(1), 116–142 (2004)
17. Katz, L.: A new status index derived from sociometric analysis. Psychometrika 18(1), 39–43 (1953)
18. Koren, Y.: Factorization meets the neighborhood: a multifaceted collaborative filtering model. In: Proc. of the 4th ACM SIGKDD int. Conf on Knowledge Discovery and Data Mining, KDD 2008, pp. 426–434. ACM, New York (2008)
19. Luo, H., Niu, C., Shen, R., Ullrich, C.: A collaborative filtering framework based on both local user similarity and global user similarity. Machine Learning 72(3), 231–245 (2008)
20. Miller, B., Albert, I., Lam, S., Konstan, J., Riedl, J.: MovieLens unplugged: experiences with an occasionally connected recommender system. In: Proc. of the 8th Int. Conf on Intelligent user Interfaces, pp. 263–266 (2003)
21. Paterek, A.: Improving regularized singular value decomposition for collaborative filtering. In: Proc. of KDD Cup and Workshop (2007)
22. Sarwar, B., Karypis, G., Konstan, J., Riedl, J.: Application of Dimensionality Reduction in Recommender System-A Case Study. Defense Technical Information Center (2000)
23. Sarwar, B., Karypis, G., Konstan, J., Riedl, J.: Item-based collaborative filtering recommendation algorithms. In: Proc. of the WWW Conf. (2001)

24. Steuer, R.E.: Multiple criteria optimization: theory, computation, and application. Wiley, New York (1986)
25. Swets, J.A.: Measuring the accuracy of diagnostic systems. Science (240), 1285–1293 (1988)
26. Takacs, G., Pilaszy, I., Nemeth, B., Tikk, D.: Investigation of various matrix factorization methods for large recommender systems. In: Proc. of the 2nd KDD Workshop on Large Scale Recommender Systems and the Netflix Prize Competition (2008)
27. Wang, F., Ma, S., Yang, L., Li, T.: Recommendation on item graphs. In: Proc. of the Sixth Int. Conf. on Data Mining, ser. ICDM 2006, pp. 1119–1123. IEEE Computer Society, Washington,DC (2006)
28. Wilson, R.C., Hancock, E.R., Luo, B.: Pattern Vectors from Algebraic Graph Theory. IEEE Trans. on Pattern Analysis and Machine Intelligence (2005)
29. Xue, G.R., Lin, C., Yang, Q., Xi, W., Zeng, H.J., Yu, Y., Chen, Z.: Scalable collaborative filtering using cluster-based smoothing. In: Proc. of SIGIR (2005)

Constraint Propagation for the Dial-a-Ride Problem with Split Loads

Samuel Deleplanque and Alain Quilliot

LIMOS, UMR CNRS 6158, Bt. ISIMA, BLAISE PASCAL university, France
{deleplan,quilliot}@isima.fr

Abstract. This paper deals with a new problem: the Dial and Ride Problem with Split Loads (DARPSL), while using randomized greedy insertion techniques together with constraint propagation techniques. Though it focuses here on the static versions of Dial and Ride, it takes into account the fact that practical DARP has to be handled according to a dynamical point of view, and even, in some case, in real time contexts. So, the kind of algorithmic solution which is proposed here, aim at making easier to bridge both points of view. First, we propose the general framework of the model and discuss the link with dynamical DARP, second, we describe the two algorithms (DARP and DARPSL), and lastly, show numerical experiments for both.

1 Introduction and Literature Review

Literature in the field of urban systems and geomatics hint a trend to a surge of new on demand flexible transportation systems (ODT): ad hoc shuttle fleets, vehicle sharing (AUTOLIB...), co-transportation (see for instance [17], [4]). This trend reflects from both environmental (climate change, overcrowded megalopolis) and economical concerns (surge of energy prices). It has also to be associated with technological advances: internet, mobile communication, geo-localization, which allow efficient monitoring of complex mobility system and large sets of heterogeneous requests.

An important Operations Research model for the management of flexible reactive transportation system is the DARP, which tries to optimize the way a given fleet of vehicles meet mobility demands emanating from people, or, in some cases from some combination of people and goods. DARP is a complex problem, which admits several formulation, most of them NP-Hard. It usually does not fit well the Integer Linear Programming framework [16] and one must try do handle it through heuristic techniques: Tabu search [7], genetic algorithms [18], partial branch/bound [16], Simulated Annealing [9], VNS techniques [20], [13], Dynamic Programming [16]-[17], Insertion techniques [6]-[12].

A basic features of DARP is that it usually derives from a dynamic context. So, algorithms for static DARP should be designed in order to take into account the fact that they will have to be adapted to dynamic and reactive context, which means synchronization mechanisms, interactions between the users and the vehicles, and uncertainty about fore coming demands.

S. Fidanova (Ed.): *Recent Advances in Computational Optimization*, SCI 470, pp. 31–50.
DOI: 10.1007/978-3-319-00410-5_3   © Springer International Publishing Switzerland 2013

In this paper, we use a generic DARP model with time windows to define the new DARP with split loads, and we propose algorithms for this model based upon randomized insertion techniques and constraint propagation. These algorithms can be easily adapted to dynamic contexts, where demand packages has to be inserted into (or eventually removed from) current vehicle schedules. This has to be done in a very short time while taking into account some probabilistic knowledge about fore coming demand packages.

The focus of the paper is to use constraint propagation on the time constraints and to allow split loads in the DARP. Little has been published on this problem, only [15]-[19]. The closest study is the Pickup and Delivery Problem with Splits Loads (PDPSL). The PDPSL solution of solved instances can give a rate up to 25% fewer vehicles used compared to the classic PDP [10]. The Tabu search is also used to solve the PDFSL [2]. For a recent review of this problem, refer to [3].

The paper is organized as follows: we first introduce the problem and discuss the link between static and dynamic formulations. Next, describe our formal DARP with split loads model, together with the performance criteria which we used. Then, we present the general insertion mechanism together with the constraint propagation techniques which we use in order to filter insertion parameters and to select the demands to be inserted. We conclude with experiments and comparison of the two resolutions.

2   The Standard Dial a Ride Problem

2.1   General Dial a Ride Problem

A Dial a Ride Problem instance is essentially defined by:

- a transit network $G = (V, E)$, which contains at least two nodes *Depot*, for the departure and the arrival, and whose arcs $e \in E$ are endowed with riding times equal to the Euclidean distance between two nodes of V, i and j, $DIST(i, j) \geq 0$, and, eventually, with other technical characteristics;
- a fleet VH of K vehicles k with a capacity CAP,
- a Demand set $D = (D_i, i \in I)$, any demand D_i being defined as a 6-uple $D_i = (o_i, d_i, D_i, F(o_i), F(d_i), Q_i)$, where:
  - $o_i \in V$ is the origin node of the demand D_i and the set of all the o_i is DE,
  - $d_i \in V$ is the destination node of the demand D_i and the set of all the d_i is AR,
  - $\Delta_i \geq 0$ is an upper bound (transit bound) on the duration of demand's processing (ride time),
  - $F(o_i)$ is a time window related to the time D_i starts being processed; $F.MIN_{o_i}$ and $F.MAX_{o_i}$ are the two bounds;
  - $F(d_i)$ is a time window related to the time D_i ends being processed,
  - Q_i is a description of the load related to D_i such that $q_{o_i} = Q_i = -q_{d_i}$,

- $\delta_j, j \in V$ is the non-negative service time necessary at the node j;
- t_i^k is the time at which the vehicle k begins service in i,
- Δ^k is the route maximum time imposed on the vehicle k.

Then, we consider in $G = (V, E)$ all the nodes corresponding to the $o_i \in V$ and $d_i \in V$ such that $V = DE \cup AR \cup \{0, 2\,|D|+1\}$ with $\{0, 2\,|D|+1\}$ the two depot nodes respectively for the departure and the arrival, $o_i \in \{1..\,|D|\}$, and $d_i \in \{(|D|+1)..(2\,|D|)\}$. Moreover we denote by ζ_j^k the total load of the vehicle k leaving the node j, $j \in V$.

Dealing with such an instance means planning the handling demands of D, by the fleet VH, while taking into account the constraints which derive from the technical characteristics of the network G, of the vehicle fleet VH, and of the 6-uples $D_i = (o_i, d_i, D_i, F(o_i), F(d_i), Q_i)$, and while optimizing some performance criterion which is usually minimizing the total distance or a mix of an economical cost (point of view of the fleet manager) and of QoS criteria (point of view of the users).

Let x_{ij}^k a boolean equals to 1 if the vehicle k travels from the node i to the node j. Then, based on [5], the mathematical formulation is the following mixed-integer program :

$$Min \sum_{k \in K} \sum_{i \in V} \sum_{j \in V} DIST(i, j) x_{ij}^k \tag{1}$$

subject to

$$\sum_{k \in K} \sum_{j \in V} x_{ij}^k = 1, \forall i \in DE \tag{2}$$

$$\sum_{j \in V} x_{ij}^k - \sum_{j \in V} x_{|D|+i,j}^k = 1, \forall i \in DE, k \in K \tag{3}$$

$$\sum_{j \in V} x_{0j}^k = 1, \forall k \in K \tag{4}$$

$$\sum_{j \in V} x_{ji}^k - \sum_{j \in V} x_{ij}^k = 1, \forall i \in DE \cup AR, k \in K \tag{5}$$

$$\sum_{i \in V} x_{i, 2|D|+1}^k = 1, \forall k \in K \tag{6}$$

$$t_j^k \geq (t_i^k + \delta_i + DIST(i, j)) x_{ij}^k, \forall i \in V, j \in V, k \in K \tag{7}$$

$$\zeta_j^k \geq (\zeta_i^k + q_j) x_{ij}, \forall i \in V, j \in V, k \in K \tag{8}$$

$$DIST(i, |(D)|+i) \leq t_{|(D)|+i}^k - (t_i^k + \delta_i) \leq \Delta_i, i \in DE \tag{9}$$

$$F.MIN_i \leq t_i^k \leq F.MAX_i, \forall i \in V, k \in K \tag{10}$$

$$t^k_{2|D|+1} - t^k_0 \leq \Delta^k, \forall k \in K \tag{11}$$

$$\zeta^k_i \leq CAP^k \tag{12}$$

$$x^k_{ij} \in \{0; 1\}, t \in R^+ \tag{13}$$

The program above is a three index formulation (report to [5] for more explanations about the objective function (1) and the constraints (2)-(13)), it exists in literature several other mathematical formulations for the DARP, even some with two index formulation [8]. But, the complexity of all these linear programs doesn't allow finding an exact solution with a solver, the operation is too time consuming. In fact, it mixes a lot of booleans and plenty of fractional numbers.

All along this work, we are going to deal with homogeneous fleets and with nominal demands, and we shall limit ourselves to static points of view but our insertion process allows flexibility for using it in a dynamic context. Still, we shall pay special attention to cases when temporal constraints are tight.

Discussion: Dynamic versus Static DARP. The problem is essentially a problem which arise in dynamic contexts, and the trend is about reactivity delays which become smaller and smaller [1]. Basically, one should consider a system which is identified by a vehicle set V, a user community C, and a supervision system S, which, because of advances in the field of geo-localization, mobile communications and remote monitoring, permanently disposes of a full knowledge about the current state of the vehicles (position, load, roadmap...) and maintains communication with both users and vehicles. All along the time, the system (centralized or decentralized) receives user request, which, in the simplest case, are characterized by a load, an origin and a destination node, and time windows related load and unload transactions, as well as about trip duration. At some instant t, supervisor S decides to launch a scheduling process P, which consider as its input the current state E of the vehicles of V, together with the currently waiting demand set D, and which, for any demand d in D, either rejects it or insert it into the current schedule of some vehicle k in V, without modifying in a significant way the way v is supposed to meet previous demands. Running P require a δ computing time, and, at time $t + \delta$, propositions are transmitted to users and updated schedules are transmitted to the vehicles, which apply them until instant t', when the whole process takes place again. Meanwhile, it may occur that some demands are dropped or that vehicles register failure (delays or user fault) [11].

In any case, one see that, in case vehicles are moving inside a small area (a urban area) and deal with a large size set of demands, process P has to insert in a fast way a demand set D into a current schedule E, and that it has to do it in a way which keeps most features of E, and preserves the ability of the system to efficiently deal with fore coming demands, that means with demands which are likely to be formulated after the instant t when P is launched. This point is the key one which motivates the approach which is going to be described here. We want an algorithmic framework which is going to be naturally compatible with this context: the use of insertion techniques is clearly going to fit the input (E, D) of the dynamic context, and the use of constraint propagation techniques is going to make easier uncertainty about fore coming demands handling.

Also, one should notice that, under this prospect, the virtual complete network which is going to be the key input data for the static model (see next section), is, in practice, going to be a dynamic network.

2.2   The Framework

The Considered Network. We treat here the general Dial a Ride Problem described above. It is known that we do not need to consider the whole transit network G = (V, E), and that we may restrict ourselves to the nodes which are either the origin or the destination of some demand, while considering that any vehicle which visits two such nodes in a consecutive way does it according to a shortest path strategy. This leads us to consider the node set $\{Depot, o_i, d_i, i \in I\}$ as made with pairwise distinct nodes, and provided with some distance function DIST, which to any pair x, y in $\{Depot, o_i, d_i, i \in I\}$, makes correspond the shortest path distance from x to y in the transit network G.

As a matter of fact, we also split the Depot node according to its arrival or departure status and to the various vehicles of the fleet VH, and we consider the input data of a Standard Dial a Ride Problem instance as defined by:

- the set $\{1..K = \mathrm{Card}(VH)\}$ of the vehicles of the homogenous fleet VH;
- the common capacity CAP of a vehicle in VH;
- the node set X = $\{DepotD(\mathrm{k}), DepotA(\mathrm{k}), \mathrm{k} = 1..K\} \cup \{o_i, d_i, i \in I\}$;
- the distance matrix DIST, whose meaning is that, for any x, y in X, DIST(x, y) is equal to the length, in the sense of the length function l, of a shortest path which connect x to y in the transit network G: we suppose that DIST, satisfies the triangle inequality.

Moreover the following characteristics, which, to any node x in X, make correspond:

- its status Status(x): Origin, Destination, $DepotA$, $DepotD$; we set $Depot = DepotD \cup DepotA$;
- its load CH(x):
  - if Status(x) $\in Depot$ then CH(x) = 0;
  - if Status(x) = Origin then CH(x) = Q_i;
  - if Status(x) = Destination then CH(x) = $-Q_i$;
- its twin node Twin(x):
  - if x = $DepotA(\mathrm{k})$ then Twin(x) = $DepotD(\mathrm{k})$ and conversely;
  - if x = o_i then Twin(x) = d_i and conversely;
- its time window F(x): for any k = 1..K, F($DepotA(\mathrm{k})$) = $[0, +\infty\ [$ = F($DepotD(\mathrm{k})$). Also, we suppose that any F(x), $x \in X$, is an interval, which may be written F(x) = [F.min(x), F.max(x)];

- its transit bound $\Delta(x)$: if x $= o_i$ or d_i, then $\Delta(x) = \Delta_i$, and $\Delta(x) = \Delta$ else, where Δ is an upper bound which is imposed on the duration of any vehicle tour.

According to this construction, we understand that the system works as follows: vehicle $k \in \{1..K\}$, starts its journey from $DepotD(k)$ at some time t$(DepotD(k))$ and ends it into $DepotA(k)$ at some time t$(DepotA(k))$, after having taken in charge some subset D(k) $=\{D_i, i \in I(k)\}$ of D: that means that for any i in I(k), vehicle k arrived in o_i at time $t(o_i) \in F(o_i)$, loaded the whole load Q_i, and kept it until it arrived in d_i at time $t(d_i) \in F(oi)$ and unloaded Q_i, in such a way that $t(d_i) - t(o_i) \leq D_i$. Clearly, solving the Standard Dial a Ride Problem instance related to those data (X, DIST, K, CAP) will mean computing the subsets D(k) $= \{D_i, i \in I(k)\}$, the routes followed by the vehicles and the time values t(x), $x \in X$, in such a way that both economical performance and quality of service be the highest possible.

Discussion: Durations and Waiting Times. Many authors include what they call service durations in their models. That means that they suppose that loading and unloading processes related to the various nodes of X require some time amount $\delta(x)$, (service time) and, so, that they distinguish, for any node $x \in X$, time values t(x) (beginning of the service) and t(x) + $\delta(x)$ (end of the service). By the same way, some authors suppose that the vehicles are always running at their maximal speed, and make a difference between the time t*(x), $x \in X$, when some vehicle arrives in x, and the time t(x) when this vehicle starts servicing the related demand (loading or unloading process). We do not do it. Taking into account service times, which tends to augment the size of the variables of the model and to make it more complex it, has really sense only if we suppose that the service times $\delta(x)$ depend on the current state (its current load) of the vehicle at the time the loading or unloading process has to be launched. Making explicitly appear waiting times t(x) - t*(x) is really useful if we make appear the speed profile as a component of the performance criterion. In case none of the situation holds, the knowledge of the routes of the vehicles and of the time value t(x), $x \in X$, is enough to check the validity of a given solution and to evaluate its performance, and then it turns out that ensuring the compatibility of the model with data which involve service times and waiting times t(x) - t*(x), $x \in X$, is only a matter of adapting the times windows F(x), the transit bounds $\Delta(x)$, $x \in X$, and the distance matrix DIST.

Discussion: The Homogeneity of the Fleet. The general case of the Dial a Ride Problem includes a homogeneous fleet of vehicles. The word "homogeneous" mean the vehicles come from (and come back to) the same depot, and have the same capacity. Our model can integrate different depots and capacities for each vehicle without changing in the framework. Moveover, $DepotD$ and $DepotA$ locations could be different because all these nodes are independent for a given route.

2.3   Modeling and Evaluation Techniques

The model described in this section needs some definitions, we set:

- First(Γ) = First element of Γ ; Last(Γ) = last element of Γ ;
- for any z in Γ:
  - Succ(Γ, z) = Successor of z in Γ ;
  - Pred(Γ, z) = Predecessor of z in Γ ;
- for any z, z' in Γ:
  - $z \ll_\Gamma z'$ if z is located before z' in Γ ;
  - $z \ll_\Gamma^= z'$ if $z \ll_\Gamma z'$ or z = z';
  - Segment(Γ, z, z') = the subsequence defined by all z in Γ such that $z \ll_\Gamma^= z \ll_\Gamma^= z'$. If z = Nil, then Segment(G, Nil, z') denotes the subsequence defined by all z in Γ such that $z \ll_\Gamma^= z'$.

In any algorithmic description, we use the symbol $\leftarrow$ in order to denote the value assignment operator: $x \leftarrow \alpha$, means that the variable x receives the value α. Thus, we only use symbol = as a comparator.

In order to provide an accurate description of the output data of our standard Dial a Ride Problem instance (X, DIST, K, CAP), we need to talk about tours and related time value sets. A tour Γ is a sequence of nodes of X, which is such that: 3

- Status(First(Γ)) = *DepotD*; Status(End(Γ)) = *DepotA*;
- For any node x in Γ, x $\neq$ First(Γ), End(Γ), Status(x) $\notin$ *Depot*;
- No node $x \in X$ appears twice in Γ ;
- For any node x = o_i (resp. d_i) which appears in Γ, the node Twin(x) is also in Γ, and we have: $x \ll_\Gamma Twin(x)$ (resp. $Twin(x) \ll_\Gamma x$).

This tour Γ is said to be load-valid iff:

- for any x in Γ, x $\neq$ First(G), we have $\sum_{y, y \ll_\Gamma x} CH(y) \leq CAP$.

Moreover, this tour Γ is said to be time-valid iff it is possible to associate, with any node x in Γ, some time value t(x), in such a way that: (E1)

- for any x in Γ, x $\neq$ Last(Γ), $t(Succ(\Gamma, x)) \geq t(x) + DIST(x, Succ(\Gamma, x))$;
- for any x in Γ, $|t(twin(x)) - t(x)| <= \Delta(x)$;
- for any x in Γ, $t(x) \in F(x)$.

In case the tour Γ is time-valid, any time value set t = {t(x), $x \in X$}, which satisfies (E1) is said to be a valid related time value set.

The tour Γ is said to be valid if it is both time valid and load valid.

For any pair (Γ, t) defined by some time-valid tour Γ and by some valid related time value set t, we may set:

- *Glob*(Γ, t) = t(End(Γ)) - t(First(Γ)): this quantity denotes the global duration of the tour Γ ;
- *Ride*(Γ, t) = $\sum_{i \in \Gamma}(t(d_i) - t(o_i))$; this quantity may be viewed as a QoS criterion, and denotes the sum of the duration of the individual trips of the demanders which are taken in charge by tour Γ ;

- $Wait(\Gamma, t) = Glob(\Gamma, t) - (\sum_{x, x \neq Last(G)} DIST(x, Succ(\Gamma, x)))$: this quantity denotes the waiting time of the vehicle involved in Γ, the waiting time
related to some node x being the time the vehicle is supposed to wait before
loading or unloading x in case it runs full speed on the route which connects
Pred(Γ, x) to x.

If A, B, C are three multi-criterion coefficients, we may define the performance
criterion $Cost_{A,B,C}$ (Γ, t) as follows: $Cost_{A,B,C}(\Gamma, t) = A.Glob(\Gamma, t) + B.Ride(\Gamma, t) + C.Wait(\Gamma, t)$.

In the experiments section we will use different criteria in order to compare
with other techniques found in literature. Our insertion techniques allow some
flexibility for this change.

So, let us suppose that we deduced from the data G = (V, E), VH = (K,
CAP), $D = (D_i = (o_i, d_i, \Delta_i, F(o_i), F(d_i), Q_i), i \in I)$, a 4-uple (X, DIST, K,
CAP), and that we are also provided with 3 multi-criterion coefficients A, B
and $C \geq 0$. Then we see that solving the related Standard Dial a Ride Problem
instance means computing:

- for any vehicle index k in 1..K, a valid tour T(k);
- a time value set $t = \{t(x), x \in X\}$;

in such a way that:

- the restriction of t to any T(k), k = 1..K, defines a valid time value set
related to T(k);
- the tour set T = {T(k), k = 1..K} induces a partition of X;
- the quantity $Perf_{A,B,C}(T, t) = \sum_{k=1..K} Cost_{A,B,C}(T(k), t)$ is the smallest
possible.

3   Constraint Propagation into an Insertion Algorithm

3.1   Handling Constraints

Let Γ a tour. The algorithm which we are going to describe in this section will
essentially be based upon the use of insertion techniques. Thus, we must be
able to check in a fast way, whether the insertion of some demand D_i inside Γ
will maintain the validity of Γ, and to get an evaluation of the quality of this
insertion. Since we want to pay a special attention to the case when temporal
constraints are tight, we are first going to provide ourselves with a package of
constraint handling tools for testing the valid tours.

First, checking the load validity of Γ is easy. In order to be able to test the
impact of the insertion of some demand into the tour Γ on the charge-validity of
this tour, we associate, with any such a tour, the quantities $\zeta(\Gamma, x)$, defined by:

- for any x in Γ, $\zeta(\Gamma, x)$ $= \sum_{y, y \ll_{\Gamma}^{=} x} CH(y)$.

Then it comes that Γ is load-valid iff for any x in Γ, $\zeta(\Gamma, x) \leq CAP$.

Second, checking the time validity of Γ, according to a current time window set FS = $\{FS(x) = [FS.min(x), FS.max(x)], x \in \Gamma\}$ may be performed through propagation of the following inference rules R_i, i = 1..5. We denote by NFact a list of nodes related to time constraints non propagated. The five inferences rules are:

Rule R1 (if (y = Succ(Γ, x))):

$$FS.min(x) + DIST(x, y) > FS.min(y)$$
$$|= FS.min(y) \leftarrow FS.min(x) + DIST(x, y); NFact \leftarrow y;$$

Rule R2 (if (y = Succ(Γ, x))):

$$FS.max(y) - DIST(x, y) < FS.max(x)$$
$$|= FS.max(x) \leftarrow FS.max(y) - DIST(x, y); NFact \leftarrow x;$$

Rule R3 (if (y = Twin(x)) and ($x \ll_\Gamma y$)):

$$FS.min(x) < FS.min(y) - \Delta(x,y)$$
$$|= FS.min(x) \leftarrow FS.min(y) - \Delta(x,y); NFact \leftarrow x;$$

Rule R4 (if (y = Twin(x)) and ($x \ll_\Gamma y$))

$$FS.max(y) > FS.max(x) + \Delta(x,y)$$
$$|= FS.max(y) \leftarrow FS.max(x) + \Delta(x,y) ; NFact \leftarrow y;$$

Rule R5 (if x $\in \Gamma$):

$$FS.min(x) > FS.max(x)$$
$$|= Fail.$$

Propagating these rules may be performed as follows:

Procedure Propagate
Input: (Γ : Tour, L: List of nodes, FS: Time windows set related to the node set of Γ);
Output: (Res: Boolean, FR: Time windows set related to node set of Γ);
Not Stop;
While L $\neq$ Nil and Not Stop do
    z $\leftarrow$ First(L); L $\leftarrow$ Tail(L);
    For i = 1..5 do Compute all the pairs (x, y) which make possible
    an application of the rule R_i and which are such that x = z or y = z;
        For any such pair (x, y) do
            Apply the rule Ri;
            If NFact is not in L then Insert NFact in L;
            If Fail then Stop;
Propagate $\leftarrow$ (Not Stop, FS);

Proposition 1
The tour Γ is time-valid according to the input time window set FS if and only if the Res component of the result of a call Propagate(FS, Γ) is equal to 1. In such a case, any valid time value set t related to Γ and FS is such that: for any x in Γ, t(x) $\in$ FS(x).

Proof
The part (only if) of the above equivalence is trivial, as well as the second part of the statement. As for the part (if), we only need to check that if we set, for any x in Γ:

– FS(x) = [FS.min(x), FS.max(x)];
– t(x) = FS.min(x);

then we get a time value set t ={t(x), x $\in$ X(Γ)} which is compatible with Γ and FS.
   End-Proof.

We denote by FP(Γ) the time window set which result from a call Propagate(Γ, L, F). FP(Γ) may be considered as the largest (in the inclusion sense) time window set which is included into F and which is stable under the rules R_i, i = 1..5, and is called the window reduction of F through Γ.

3.2   Evaluating a Tour

Let us consider now the tour Γ, provided with the window reduction set FP(Γ). We want to get some fast estimation of the best possible value $Cost_{A,B,C}(\Gamma, t) = A.Glob(\Gamma, t) + B.Ride(\Gamma, t) + C.Wait(\Gamma, t), t \in Valid(\Gamma)$. We already noticed that it could be done through linear programming or through general shortest path and circuit cancelling techniques. Still, since we want to perform this evaluation process in a fast way, we design two ad hoc procedures EVAL1 and EVAL2:

– the EVAL1 procedure works in a greedy way, by first assigning to the node First(Γ) its largest possible time value, and by next performing a Bellman process in order to assign to every node x in Γ its smallest possible time value.
– the EVAL2 procedure starts from a solution produced by EVAL1, and improves it by performing a sequence of local moves, each move involving a single value t(x), $x \in \Gamma$.

Γ being some valid tour, we denote by VAL1(Γ) and VAL2(Γ) the values respectively produced by the application of EVAL1 and EVAL2 to Γ.

3.3   The Insertion Mechanism

The insertion process works in a very natural way. Let Γ be some valid tour, let $D_i=(o_i, d_i, \Delta_i, F(o_i), F(d_i), Q_i)$ be some demand whose origin and destination nodes are not in Γ, and let x, y be two nodes in Γ, such that $x \ll_{\Gamma}^{=} y$. Then we denote by INSERT(Γ, x, y, i) the tour which is obtained by:

- locating o_i between x and Succ(Γ, x);
- locating d_i between y and Succ(Γ, y).

We say that the tour **INSERT(Γ, x, y, i)** results from the insertion of demand D_i into the tour Γ according to the insertion nodes x and y. The tour INSERT(Γ, x, y, i) may not be valid. So, before anything else, we must detail the way the validity of this tour is likely to be tested.

Testing the Load-Admissibility of INSERT(Γ, x, y, i)

We only need to check, that for any z in Segment(Γ, x, y) = {z such that $x \ll_{\Gamma}^{=} z \ll_{\Gamma}^{=} y$} we have, $\zeta(\Gamma, x) + Q_i \leq$ CAP. It comes that we may set:

Procedure Test-Load(Γ, x, y, i):
Test-Load $\leftarrow$ {For any z in Segment(Γ, x, y), $\zeta(\Gamma, x) + Q_i \leq$ CAP};

Testing the Time-Admissibility of INSERT(Γ, x, y, i)

It should be sufficient perform a call Propagate(Γ, {o_i, d_i}, FP(Γ)), while using the list {o_i, d_i} as a starting list. Still, such a call is likely to be time consuming. So, in order to make the testing process go faster, we introduce several intermediary tests, which aim at interrupting the testing process in case non-feasibility can be easily noticed:

- the first test Test-Node aims at checking the feasibility of the insertion of a node u, related to some load Q, between two consecutive node z and z' of a given tour Γ. It only provides us with a necessary condition for the feasibility of this insertion.
- the second test Test-Node1 aims at checking the feasibility of the insertion of an origin/destination node u, v, related to some load Q, between two consecutive node z and z' of a given tour Γ (e.g. into an empty tour). Again, it only provides us with a necessary condition for the feasibility of this insertion.

So, testing the admissibility of a tour INSERT(Γ, x, y, i) may be performed through the following procedure:

Procedure Test-Insert(Γ, x, y, i): (Test: Boolean, Val: Number);
If $x \neq y$ then
  Test $\leftarrow$ Test-Node(Γ, x, Succ(Γ, x), o_i, Q_i) $\wedge$ Test-Node(Γ, y, Succ(Γ, y), d_i, Q_i);
Else Test $\leftarrow$ Test-Node1(Γ, x, Succ(Γ, x), o_i, d_i, Q_i);
If Test = 1 then Test $\leftarrow$ Test-Load(Γ, x, y, i);
  If Test = 1 then (Test, F1) $\leftarrow$ Propagate(Γ, {o_i, d_i }, FP(Γ);
    If Test = 1 then Val $\leftarrow$ EVAL1(INSERT(Γ, x, y, i), F1).Val;
Else Val $\leftarrow$ Undefined;
Test-Insert $\leftarrow$ (Test, Val - Val1(Γ));

3.4   The Insertion Process

So, this process takes as input the demand set $D = (D_i=(o_i, d_i, \Delta_i, F(o_i), F(d_i), Q_i))$, $i \in I$), the 4-uple (X, DIST, K, CAP), and 3 multi-criteria coefficients A, B and C ≥ 0, and it works in a greedy way through successive insertions of the various demands $D_i = (o_i, d_i, \Delta_i, F(o_i), F(d_i), Q_i)$ of the demand set D. The basic point is that, since we are concerned with tightly constrained time windows and transit bounds, we use, while designing the INSERTION algorithm, several constraint propagations tricks. Namely, we make in such a way that, at any time we enter the main loop of this algorithm, we are provided with:

- the set $I_1 \subset I$ of the demands which have already been inserted into some tour T(k), k = 1..K;
- current tours T(k), k = 1..K: for any such a tour T(k), we know the related time windows FP(T(k))(x), $x \in T(k)$, as well as the load values $\zeta(T(k), x), x \in T(k)$, and the values VAL1(T(k)) and VAL2(T(k));
- the knowledge, for any i in J = (I - I1) of the set $U_{free}(i)$ of all the 4-uple (k, x, y, v), k = 1..K, $x, y \in T(k)$, $v \in Q$, such that a call Test-Insert(T(k), x, y, i) yields a result (1, v). We denote by $N_{free}(i)$ the cardinality of the set $K_{free} = \{k = 1..K$, such that there exists a 4-uple (k, x, y, v) in $U_{free}(i)\}$: $N_{free}(i)$ provides us with the number of vehicles which are still able to deal with demand D_i.

Then, the INSERTION algorithm works according to the following scheme:

- First, it picks up some demand i_0 in J, among those demands which are the most constrained, that means which are such that $N_{free}(i_0)$ is small: more specifically, if there exists i such that $N_{free}(i) = 1$, then i_0 is chosen in a random way among those demand indices i in J which are such that $N_{free}(i) = 1$; else we select randomly in a set of demands j with $N_{free}(j)$ inside {2, $N_{freeMAX}$ }. $N_{freeMAX}$ becomes a parameter of the INSERTION.   (E2)
- Next, it picks up (k_0, x_0, y_0, v_0) in $U_{free}(i_0)$ which simultaneously corresponds to one of the smallest values v, and to one of the smallest values EVAL2(INSERT(T(k), x, y, i_0)).Val - VAL2(T(k)): more specifically it first builds the list L-Candidate of the N_1 (up to five) 4-uples (k, x, y, v) in $U_{free}(i_0)$ with best (smallest value v). For any such a 4-uple, it computes the value w = EVAL2(INSERT(T(k), x, y, i_0)).Val - VAL2(T(k)), and it orders L-Candidate according to increasing values w. Then it randomly chooses (k_0, x_0, y_0, v_0) among those N2 $\leq$ N1 first 4-uples in L-Candidate. N_1 and N_2 become two parameters of the INSERTION procedure.   (E3)
- Next it inserts the demand i_0 into T(k_0) according to the insertion nodes x_0, y_0, which means that it replaces T(k_0) by INSERT(T(k_0), x_0, y_0, i_0);
- Next it defines, for any $i \in J$, the set $\Lambda(i)$ as being the set of all pairs (x, y) such that there exists some 4-uple (k_0, x', y', v) in $U_{free}(i)$, which satisfies:
  - $(x' = x)$ or (($x' = x_0$) and $x' = \text{Pred}(T(k_0), x))$ or (($x' = x_0 = y_0$) and ($x' = \text{Pred}(\text{Pred}(T(k_0),x))))$;
  - $(y' = y)$ or (($y' = y_0$) and $y' = \text{Pred}(T(k_0), y))$ or (($y' = x_0 = y_0$) and $(y' = \text{Pred}(\text{Pred}(T(k_0),y))))$;   (E4)

– Finally, it performs, for any pair (x, y) in $\Lambda(i)$, a call Test-Insert($T(k_0)$, x, y, i), and it updates $U_{free}(i)$ and $N_{free}(i)$ consequently.

This can be summarized as follows:

Procedure INSERTION(N_1 and N_2 : Integer): (T: tour set, t: time value set, Perf: induced $Perf_{A,B,C}(T, t)$ value, Reject: rejected demand set);
    For any k = 1..K do
        T(k) $\leftarrow$ {$DepotD$(k), $DepotA$(k)};
        t($DepotD$(k)) = t($DepotA$(k)) $\leftarrow$ 0;
    I1 $\leftarrow$ Nil ; J $\leftarrow$ I ; Reject $\leftarrow$ Nil;
    For any $i \in J$ do
        N_{free}(i) $\leftarrow$ K;
        U_{free}(i) $\leftarrow$ all the possible 4-uple (k, x, y, v), $k \in K$, x, y $\in$ {$DepotD$(k), $DepotA$(k)}, $x \ll_{T(k)} y$, v = **EVAL2**({$DepotD$(k), o_i, d_i, $DepotA$(k)}).Val;
    While $J \neq Nil$ do
        Pick up some demand i_0 in J as in (E2); Remove i_0 from J;
        If $U_{free}(i_0)$ = Nil then *Reject* $\leftarrow$ *Reject* $\cup$ {$i0$};
        Else
            Derive from $U_{free}(i_0)$ the L-Candidate list and pick up (k_0, x_0, y_0, v_0) in L-Candidate as in (E3);
            $T(k_0) \leftarrow$ **INSERT**($T(k_0), x_0, y_0, i_0$);
            $\delta \leftarrow$ **EVAL2**($T(k_0)$).δ; Insert i_0 into I_1 ;
            For any x in $T(k_0)$ do t(x) $\leftarrow \delta(x)$;
            For any $i \in J$ do
                $\Lambda(i) \leftarrow$ {all pairs (x, y) such that there exists some 4-uple (k_0, x', y', v) in $U_{free}(i)$, which satisfies (E4);
                For any pair (x, y) in $\Lambda(i)$ do
                    (Test, Val) $\leftarrow$ **Test-Insert**($T(k_0)$, x, y, i);
                    Remove (k_0, x, y, v) from $U_{free}(i)$ in case such a 4-uple exists and update $N_{free}(i)$ consequently;
                    If Test = 1 then insert (k_0, x, y, Val) into $U_{free}(i)$ and update $N_{free}(i)$ consequently;
    Perf $\leftarrow$ $Perf_{A,B,C}$ (T, t);
    INSERTION $\leftarrow$ (T, t, Perf, Reject);

Since the above (I1) and (I2) instruction may be written in a non deterministic way, the whole INSERTION algorithm becomes non deterministic and may be used inside some MONTE-CARLO framework:

RANDOM-INSERTION(N_1, N_2, P: Integer) Scheme;
    For p = 1..P do
    Apply the INSERTION(N_1, N_2) procedure;

Keep the best result (the pair (T, t) such that |Reject| is the smallest possible, and which is such that, among those pairs which minimize |Reject|, it yields the best $Perf_{A,B,C}$ (T, t) value).

4   Dial-a-Ride Problem with Split Loads

4.1   Model and Framework updated

The Dial-a-ride problems with split loads means we allow related to some demand to be split in several pieces and to transported separately. Such a situation may occur in the case of good transportation (large scale load management) as well as in the case of people transportation (group management). Difficulties start with modeling, since the way loads Q_i are divided into load-pieces $Q_{i,j}$, $j = 1..n(i)$, is part of the problem.

We based on the general Dial-a-ride Problem defined above and we update :

- the set X which gives rise to a infinite set $Z = Z(X)$, which derives from X by replacing every node x such that Status(x) = {Origin, Destination}, by nodes (x, s), $s \in N$. This splitting process will allow us to distinguish the nodes of X which are related to some demand D_i according to the load-pieces $Q_{i,j}$, j = 1..n(i): the meaning of node (o_i, s) is that if a tour T(k) contains this node (o_i, s), then it will also contain the node (d_i, s), and vehicle k will ensure the transportation of some load-piece $Q_{i,j}$ from o_i to d_i.
- the DIST matrix which may be considered as extended in a natural way as a DIST function which is defined on Z.Z;
- the Twin function : for any node $z = (x, s) = (o_i, s)$ $((d_i, s))$ which appears in Γ, the node Twin(z) = (d_i, s) $((o_i, s))$ is also in Γ, and we have: z << Twin(z) ((Twin(z) << z));
- the ride is computed by the duration between the first time's origin node and the last time's destination node for a given demand;
- the maximum ride time could be considered in two different ways (for a given demand) :
  - it bounds the duration given by the first time's origin node and the last time's destination node,
  - each load-pieces is independent and bounded by the same maximum ride time (like in our experiments).

So, the Dial-a-Ride Problem with split loads may be put in a formal way as follows:

The Dial-a-Ride Problem with Split Loads
Input: the demand set D = $(D_i = (o_i, d_i, \Delta_i, F(o_i), F(d_i), Q_i), i \in I)$, the 4-uple (X, DIST, K, CAP) which we defined above, and 3 multi-criteria coefficients A, B and $C \geq 0$;
Output: a triple (T, t, Q) where T = {T(k), $k \in K$} is a time valid tour family, Q = {Q(k), $k \in K$} is a family of related valid load value sets Q(k) = {Q(k)(z), $z \in T(k)$}, and t = {t(k), $k \in K$} is a family of related valid time value sets t(k) = {t(k)(z), $z \in T(k)$} such that:

- for every $i \in I$, we have:
  $$\sum_{k=1..K} \sum_{(i,s) \text{ if } o_i \in T(k)} Q(k)(o_i, s) = -\sum_{k=1..K} \sum_{(i,s) \text{ if } d_i \in T(k)} Q(k)(d_i, s)$$

– The quantity $Perf_{A,B,C}(T, t, Q) = \sum_{k=1..K} Cost_{A,B,C}(T(k), Q(k), t(k))$ is the smallest possible.

Active set related to a feasible triple (T, t, Q): it is the set of the active nodes (x, s) in $Z = Z(X)$, which means the nodes which belong to some tour T(k), k = 1..K. The general algorithmic scheme **INSERTION-SPLIT-LOADS** will come as follows:

Initialize (T, t, Q); Initialize the sets $U_{free}(i)$, $i \in I$;
Initialize the active Z-ACT: Z-ACT <- Nil;
J <- I; For any i in J, set $Q\text{-}Aux_i$ <- Q_i; Reject <- Nil;
While $J \neq Nil$ do
    Picks up some demand i_0 in J; Remove i_0 from J;
    If $U_{free}(i_0)$ = Nil then Reject <- Reject $\cup \{i_0, Q\text{-}Aux_{i_0}\}$;
    Else
        Compute s_0; Create two new active nodes (o_{i_0}, s_0) and (d_{i_0}, s_0) and insert them into Z-ACT;
        Derive from $U_{free}(i_0)$ a L-Candidate list;
        Pick up $(k_0, x_0, y_0, v_0, Q(k_0)(o_{i_0}, s_0))$ in L-Candidate;   (E5)
        $T(k_0)$ <- INSERT($T(k_0)$, x_0, y_0, i_0);
        Update t and Q;
        Update the sets $U_{free}(i)$, $i \in J$;
        If $Q(k_0)(o_{i_0}, s_0)$ = $Q\text{-}Aux_{i_0}$ then Remove i_0 from J else replace $Q\text{-}Aux_{i_0}$ by $Q\text{-}Aux_{i_0}$ - $Q(k_0)(o_{i_0}, s_0)$;

4.2   Trade-Off between Load and Speed: The Load-Distribute Problem

As a matter of fact, performing Instruction (E5) above, which means conveniently the parameters k_0, x_0, y_0 and $Q(k_0)(o_{i_0}, s_0)$ of the insertion process, also means defining some trade-off between the value $Q(k_0)(o_{i_0}, s_0)$, which we would like to be the larger possible, and the quality of the insertion in relation to the criterion measure $Perf_{A,B,C}(T, t)$ and to the values which are returned by EVAL1 and EVAL2. In order to define this trade-off, we do not exactly follow the above algorithmic scheme: instead, we proceed in a specific way, which consists in handling the whole demand D_{i_0} inside a same iteration, while eventually splitting into several blocks and simultaneously distributing those blocks in an ad hoc way between the different tours. In order to put it in a more precise way, let us suppose that we are dealing as above with a demand index i_0 in J, in such a way that $Q\text{-}Aux_{i_0} = Q_{i_0}$ and with a L-Candidate list which we derived from the set $U_{free}(i_0)$. The elements of L-Candidate are 5-uple (k, x, y, v, q), which express the feasibility of the transportation by vehicle k of a load q from o_{i_0} to d_{i_0}, respectively inserted into the tour T(k) between x and Succ(T(k), x) and between y and Succ(T(k), y), the number v providing us with the EVAL2 value of this insertion. Then we try to solve the following problem:

Load-Distribute Problem

{Select a collection $\Lambda = \{(k_1, x_1, y_1, v_1, q_1),.., (k_s, x_s, y_s, v_s, q_s)\}$ of 5-uples of L-Candidate, in such a way that:

- the k_j, j = 1..s, are pairwise distincts;
- $\sum_{j=1..s} q_j \geq Q_{i_0}$;
- $\sum_{j=1..s} v_j$ is the smallest possible}

While this problem may be easily solved in an exact way through a bipartite graph matching procedure, we deal with it in a fast way through a simple heuristic Load-Distribute procedure. In case we don't find any feasible solution to this problem, then we reject the whole demand i_0. Else, we create the active nodes (o_{i_0}, j), (d_{i_0}, j), j = 1..s , we add them to Z, and, for every index value j = 1..s, we replace the tour $T(k_j)$ by the tour INSERT($T(k_j)$, x_j, y_j, i_0), and we consequently update the time value set $t(k_i)$ and the load value set $Q(k_i)$.

Remark 1. The failure of the Load-Distribute test does not completely mean that the insertion of demand i_0 cannot be performed: theoretically, one might build instances which would make a distributed insertion possible, under the condition that a same vehicle is going to support several distinct nodes (o_{i_0}, j). In such a case, we should perform the insertion of a first part of demand i_0, and next try again with the remaining part, while eventually using the same vehicle as for the first part. Still, practically, such a configuration is likely to occur very scarcely, and, so, we decide not to take it into account.

5   Computational Experiments

5.1   Experiment on the Classic Dial a Ride Problem

This first experiment deals with the two sets of instances defined in [5]. We integrated the same mono criterion objective function given by Cordeau: the

Table 1. Resolution of the set a [5]

Inst.	Lb	Opti	Ub	c1(s)	TI	Gap	Wait	Ride	c2(s)
a2-16	294.25	294.25	294.25	1.1	294.25	0.00	387.32	344.54	0.0
a2-20	344.83	344.83	344.83	2.6	344.83	0.00	605.44	455.32	0.1
a2-24	431.12	431.12	431.12	8.5	431.12	0.00	536.79	603.31	0.3
a3-18	300.48	300.48	300.48	4.6	300.81	0.11	196.65	419.35	0.7
a3-24	344.83	344.83	347.42	7.6	344.83	0.00	642.72	628.86	1.5
a3-30	494.85	494.85	494.85	9.8	495.26	0.08	721.21	732.86	16.3
a3-36	583.19	583.19	584.44	105.1	589.86	1.14	868.83	903.77	13.8
a4-16	282.68	282.68	282.68	5.6	283.10	0.15	100.72	307.00	0.3
a4-24	375.02	375.02	378.13	5.6	376.21	0.32	527.81	581.60	94.0
a4-32	485.5	485.5	487.81	30.7	487.10	0.33	593.75	796.45	29.4
a4-40	557.69	557.69	582.26	8328.5	565.95	1.42	1112.33	824.32	63.3
a4-48	668.82	NA	709.47	14542.6	700.30	NA	966.85	1132.92	30.8

minimization of the total distance. The instances have between 16 and 48 requests which have to be supported by a fleet of 2 to 4 vehicles, and have been divided into subsets a and b. In the first set, $CAP = 3$, the loads are all equal to 1, and the maximum riding time is 30min. For the second set, $CAP = 6$, the load q is randomly chosen according to a uniform distribution such as $q = 1..CAP$, and the maximum riding time is 40min. All the demands are randomly chosen in the square [-10,10].[-10,10] according to a uniform distribution, and all the routing costs between two nodes are equal to the Euclidean distance. The heuristics proposed in this paper was implemented in C++ and compiled with GNU GCC. Each replication was run on the same thread of an Intel Q8300 (2.5 GHz).

Table 1 shows the results obtained for the a first set of instances, and table 2 gives the results for the b second set. Lb, Ub, $Opti$, and TI are the best lower bound, the best upper bound, the known optimal value ([5]-[14]), and the result obtained with our insertion techniques respectively. The cpu times are in seconds for the first table and in minutes for the second table. $c1$ is the literature best cpu time and $c2$ is the cpu time obtained in our experiment. gap is the gap in percentage between the optimal distance and our result.

Almost each time, our heuristic found the optimal solution known in the literature and the worst gap obtained was 2,38%. We obtained these results quickly, the cpu times are low compared to previous studies, and a good solution is obtained in little time. Also, we show that our solution can be used with other objective functions, proving one the flexibility aspects of our solution.

Table 2. Resolution of the set b [5]

Inst.	Lb	Opti	Ub	$c1(m)$	TI	Gap	Wait	Ride	$c2(m)$
b2-16	309.41	309.41	309.61	0.2	309.41	0.00	386.45	448.66	0.7
b2-20	332.64	332.64	334.93	0.0	332.64	0.00	458.17	465.23	0.6
b2-24	444.71	444.71	445.11	0.1	444.71	0.00	475.88	674.12	2.6
b3-18	301.64	301.64	301.8	0.7	301.65	0.00	278.70	479.38	0.7
b3-24	394.51	394.51	394.57	3.6	397.47	0.75	609.62	572.61	3.5
b3-30	531.44	531.44	536.04	6.8	534.23	0.52	785.71	857.28	3.2
b3-36	603.79	603.79	611.79	62.1	603.79	0.00	919.58	942.81	0.9
b4-16	296.96	296.96	299.07	0.8	296.96	0.00	218.97	402.16	2.8
b4-24	371.41	371.41	380.27	5.9	371.41	0.00	490.06	567.75	0.1
b4-32	494.82	494.82	500.92	176.8	506.60	2.38	921.55	749.85	1.9
b4-40	591.76	656.6	662.91	240.0	662.74	0.94	1013.20	1021.47	3.5
b4-48	586.91	673.8	685.46	240.0	684.83	1.64	1458.76	1262.59	5.2

5.2   Experiment on the Dial a Ride Problem with Split Loads

In this section we present our instances generated in the same square [-10,10].[-10,10] as above in order to test our heuristic that solves the DARPSL. In all demands, the *origin* time window is tight (15 minutes) and the *destination* time window is large. Moreover, their *origin* location are randomly located in the rectangle [-10,-9].[-10,10], and their *destination* location is generated in the square

[9,10].[-0.5,0.5]. The depot point is located on the center of the main square. We generated four sets of 10 instances (20 to 65 demands managed by 4 to 10 vehicles). All the random generation has been computed by a uniform distribution.

Table 3. Instances solved by the DARP's heuristic

K	$|D|$	Glob	Ride	Dist	R_{Succ}	R_{Insert}	cpu(s)
4	20	836.3	953.8	416.9	71.2	98.1	0.3
6	35	1366.7	1590.7	714.2	58.0	98.0	0.7
8	50	1831.0	2203.1	1038.4	28.4	96.3	1.4
10	65	2349.7	2758.5	1347.2	25.2	95.8	2.4
Av.		1595.9	1876.5	879.2	45.7	97.0	1.2

The optimization uses a mono criterion : the minimization of the total distance. Our two algorithms proposed in this paper were applied to the four sets, each instance has been solved with 100 runs. Table 3 and table 4 report the results obtained with the classic problem and the problem with split loads respectively. *Glob* and *Ride* are the times reported from the best run, *Dist* is the best total distance obtained, R_{Succ} (%) is the insertion average rate (over the runs) of all the demands, R_{Insert} (%) the average rate of insertion for each demand, and *cpu(s)* is the time in seconds for the 100 runs.

Table 4. Instances solved by the DARPSL's heuristic

K	$|D|$	Glob	Ride	Dist	R_{Part}	R_{Succ}	R_{Insert}	cpu(s)
4	20	856.3	1113.1	356.8	1.205	93.2	99.6	0.5
6	35	1269.2	1787.2	605.5	1.216	78.4	98.8	1.3
8	50	1689.7	2542.1	850.2	1.216	53.6	97.8	2.6
10	65	2024.7	2981.0	1100.3	1.220	47.6	97.1	4.2
Av.		1460.0	2105.9	728.2	1.214	68.2	98.3	2.1

We report the average number of divisions per demand (R_{Part}). So, for the four sets, this rate is 1,214 (each demand is divided by 1,214 on average).

We observed that the split loads gives us better solutions, we obtained distances 20% less than the other problem. *Glob* also decreased (the average number of vehicles leaving the depot is lower than the classic problem). *Ride* increased because it is computed on the difference between the last date of the pickups at the destination point and the first date of the pickups at the origin point.

6   Conclusion

The static multi-vehicle DARP with Time Windows requires approximate solutions in order to be solved in a reasonable time. We have described an implementation of some insertion techniques using constraint propagation. This solution makes it possible to obtain good results in little time. In addition, we formulate an objective function which optimizes the combination of QoS and cost's minimization. But, in order to compare with tests found in literature, we prove the flexibility of our framework by changing the objective function without modification of the framework itself. Despite this change, we show that our solution is effective.

We also propose a new problem: the Dial a Ride Problem with split loads. This problem gives us better solving method compared to the classic DARP, we get 20% shorter routes with the resolution of the DARPSL with our instances.

In a future work, we could solve instances in real context, and improve our solution by integrating *inserability* demand calculator.

Acknowledgments. We wish to thank you the Conseil Regional d'Auvergne and the FEDER of the European Union.

References

[1] Ghiani, G., Laporte, G., Attanasio, A., Cordeau, J.F.: Parallel tabu search heuristics for the dynamic multi-vehicle dial-a-ride problem. Parallel Computing 30(3), 377–387 (2004)

[2] Hertz, A., Archetti, C., Speranza, M.G.: A tabu search algorithm for the split delivery vehicle routing problem. Transportation Science 40(1), 64–73 (2006)

[3] Speranza, M.G., Archetti, C.: The split delivery vehicle routing problem: A survey. In: The Vehicle Routing Problem: Latest Advances and New Challenges, pp. 103–122. Springer, US (2008)

[4] Chevrier, R.: Optimisation de transport à la demande dans des territoires polarisés. PhD. Thesis. Université d'Avignon et des Pays de Vaucluse, 242p (2008)

[5] Cordeau, J.-F.: A branch-and-cut algorithm for the dial-a-ride. Operation Research 54(3), 573–586 (2006)

[6] Jaw, J., Odoni, A., Psaraftis, H., Wilson, N.: A heuristic algorithm for the multi-vehicle many-to-many advance request dial-a-ride problem. Transportation Research B 20B, 243–257 (1986)

[7] Laporte, G., Cordeau, J.-F.: A tabu search heuristic algorithm for the static multi-vehicle dial-a-ride problem. Transportation Research B 37, 579–594 (2003)

[8] Laporte, G., Cordeau, J.F.: The dial-a-ride problem: models and algorithms. Annals of Operations Research (2007)

[9] Stone, J.R., Baugh Jr., J.W., Kakivaya, D.K.R.: Intractability of the dial-a-ride problem and a multiobjective solution using simulated annealing. Engineering Optimization 30(2), 91–124 (1998)

[10] Trudeau, P., Dror, M.: Savings by split delivery routing. Transportation Science 23(2), 141–145 (1989)

[11] Schrijver, A., Grötschel, M., Lovász, L.: Geometric algorithms and combinatorial optimization. Springer (1988)

[12] Rygaard, J., Madsen, O., Ravn, H.: A heuristic algorithm for the a dial-a-ride problem with time windows, multiple capacities, and multiple objectives. Annals of Operations Research 60, 193–208 (1995)

[13] Moll, R., Healy, P.: A new extension of local search applied to the dial-a-ride problem. European Journal of Operational Research 83, 83–104 (1995)

[14] Parragh, S.: Introducing heterogeneous users and vehicles into models and algorithms for the dial-a-ride problem. Transportation Research Part C: Emerging Technologies 19(5), 912–930 (2011)

[15] Parragh, S.N.: Solving the dial-a-ride problem with split requests and profits. CO 2012 (2012)

[16] Psaraftis, H.: An exact algorithm for the single vehicle many-to-many dial-a-ride problem with time windows. Transportation Science 17, 351–357 (1983)

[17] Chatonnay, P., Josselin, D., Chevrier, R., Canalda, P.: Comparison of three algorithms for solving the convergent demand responsive transportation problem. In: ITSC 2006, 9th Int. IEEE Conf. on IntelligentTransportation Systems, Toronto, Canada, pp. 1096–1101 (2006)

[18] Bergvinsdottir, K.B., Jorgensen, R.M., Larsen, J.: Solving the dial-a-ride problem using genetic algorithms. Journal of the Operational Research Society 58(10), 1321–1331 (2007)

[19] Quilliot, A., Deleplanque, S.: Dial a ride problem avec transbordement et division du chargement. 14e conférence ROADEF. 14-15-15 Février (résumé accepté, 2013)

[20] Hartl, R.F., Parragh, S.N., Doerner, K.F.: Variable neighborhood search for the dial-a-ride problem. Computers & Operations Research 37, 1129–1138 (2010)

ACO and GA for Parameter Settings of *E. coli* Fed-Batch Cultivation Model

Stefka Fidanova[1], Olympia Roeva[2], and Maria Ganzha[3]

[1] IICT-Bulgarian Academy of Science,
Acad. G. Bonchev Str., bl. 25A,
1113 Sofia, Bulgaria
stafka@parallel.bas.bg
[2] IBFBMI-Bulgarian Academy of Science,
Acad. G. Bonchev Str., bl.105,
1113 Sofia, Bulgaria
olympia@biomed.bas.bg
[3] System Research Institute,
Polish Academy of Sciences,
Newelska Str. 6, 01-447 Warsaw, Poland
maria.ganzha@ibspan.waw.pl

Abstract. *E. coli* plays significant role in modern biological engineering and industrial microbiology. In this paper the Ant Colony Optimization algorithm and Genetic algorithm are proposed for parameter identification of an *E. coli* fed-batch cultivation process model. A system of nonlinear ordinary differential equations is used to model the biomass growth and the substrate utilization. We use real experimental data set from an *E. coli* MC4110 fed-batch cultivation process for performing parameter optimization. The objective function was formulated as a distance between the model predicted and the experimental data. Two different distances were used and compared – the Least Square Regression and the Hausdorff Distance. The Hausdorff Distance was used for the first time to solve the considered parameter optimization problem. The results showed that better results concerning model accuracy are obtained using the objective function with a Hausdorff Distance between the modeled and the measured data. Although the Hausdorff Distance is more time consuming than the Least Square Distance, this metric is more realistic for the considered problem.

Keywords: ant colony optimization, genetic algorithm, least square distance, Hausdorff distance.

1 Introduction

A lot of proteins are produced by the modified genetically microorganisms. One of the most used host organisms in the process is the *Escherichia coli* [47]. Furthermore, the *E. coli* is still the most important host organism for the recombinant

S. Fidanova (Ed.): *Recent Advances in Computational Optimization*, SCI 470, pp. 51–71.
DOI: 10.1007/978-3-319-00410-5_4   © Springer International Publishing Switzerland 2013

protein production. In many cases, cultivation of recombinant micro-organisms e.g. the *E. coli*, is the only economical way to produce pharmaceutical biochemicals such as: interleukins, insulin, interferons, enzymes and growth factors, etc. Simple bacteria, like the *E. coli*, are manipulated to produce these chemicals so that they are easily harvested in vast quantities for use in medicine. Scientists may know more about the *E. coli* than they do know about any other species on earth. Research on the *E. coli* accelerated after 1997, after publication of its entire genome.The scientists were able to survey all 4,288 of its genes, discovering how groups of them worked together to break down food, make new copies of the DNA and do other tasks. However, despite decades of research, there rest a lot more to know about the *E. coli*. In 2002, they formed the *International E-coli Alliance*, for organization of projects that many laboratories could work together. As knowledge of the *E. coli* grows, scientists are starting to build models of the microbe that capture some of its behavior. It is important to be able to simulate how fast the microbe will grow on various sources of food, and how its growth changes if individual genes are knocked out. These questions are best answered by application of mathematical modeling.

Modeling of biotechnological processes is a common tool in process technology. It is obvious that the model is always a simplification of the reality. This is especially true when trying to model natural systems containing living organisms. However, for many industrial relevant processes, detailed models are not available due to the insufficient understanding of the underlying phenomena. These models can be too complicated and/or impossible to be solved. Therefore the specialists try to separate the most important components, and to create simplified models, which are as close as possible to the real processes. The mathematical models are very useful and effective tools in describing those effects. They are of great importance for control, optimization, or for understanding of the process. Thus the numerical solution of the models is fundamental for the development of powerful, though economical, methods in the fields of bioprocess design, plant design, scale-up, optimization and bioprocess control [40,30]. Some of the recent researches and developed models of the *E. coli* were presented in [10,21,22,27,31,42].

A common approach to model cellular dynamics is by systems of nonlinear differential equations. Obviously, parameter identification of a nonlinear dynamic model is more difficult than the linear one, as no general analytic results exist. The difficulties that may arise are such as: convergence to local solutions if standard local methods are used, over-determined models, badly scaled model function, etc. The problem is NP-hard and it is unpractical to be solved with exact or traditional numerical method. Therefore, existing research results indicate that the most useful solution method is by application of some metaheuristics. During the last decade metaheuristic techniques have been applied in a variety of areas. Heuristics can obtain suboptimal solution in ordinary situations and optimal solution in particular cases. Since the considered problem has been known to be NP-complete, using heuristic techniques can solve this problem more efficiently. Three best known (and most studied) heuristic approaches are: the iterative improvement algorithms, the probabilistic optimization algorithms, and the constructive heuristics. In this context, the evolutionary algorithms like: (a) Genetic Algorithms (GA) [18,19,25], (b) Evolution Strategies, (c) Ant Colony Optimization (ACO) [12,13,14,17], (d) Particle Swarm Optimization [46], (e) Tabu Search (TS) [49], (f) Simulated Annealing (SA) [23], (g) estimation of distribution algorithms, (h) scatter search, (i) path relinking, (j) greedy randomized adaptive search procedure, (k) multi-start and iterated local search, (l) guided local search, and (m) variable neighborhood search are - among others - often listed as examples of classical metaheuristics [6,44,45].

Obviously, they all have individual historical backgrounds and follow different paradigms and philosophies [7]. In this work the ACO and GA are chosen as the most common direct methods used for the global optimization.

The ACO is a rapidly growing research area of population-based metaheuristics that can be used to find approximate solutions to difficult optimization problems. It is applicable for a broad range of optimization problems, can be used in dynamic applications (adapts to changes such as new distances, etc.) and in some complex biological problems [15,16,41]. Recall that the ACO can compete with other global optimization techniques like GAs and SA. Overall, the ACO algorithms have been inspired by the real-world ant behavior. In nature, ants usually wander randomly, and upon finding food return to their nest while laying down pheromone trails. If other ants find such a path, they are likely to not continue traveling at random, but to follow the trail instead, returning and reinforcing it (if they eventually find food). However, as time passes, the pheromone starts to evaporate. Therefore, the more time it takes for an ant to travel down the path and back again, the more time the pheromone has to evaporate and the path becomes less noticeable. A shorter path, in comparison, will be visited by more ants and thus the pheromone density remains high for a longer time. The ACO is usually implemented as a team of intelligent agents which simulate the ants behavior, walking around the graph representing the problem to solve using mechanisms of cooperation and adaptation.

The GA is one of the oldest and well learned metaheuristics. The idea for it comes from the Darwinian theory for evolution. The two parents combines in a random way. If the ancestors are better than the parents they survive and if they are worst they will dye. Thus during the time the population improves.

In this paper the ACO and GA are applied for parameter identification of a system of the *E. coli* fed-batch cultivation process, described in terms of a mathematical model. Specifically, a system of nonlinear ordinary differential equations is proposed to model the *E. coli* biomass growth and substrate (glucose) utilization. The parameter optimization is performed using real experimental data set from the *E. coli* MC4110 fed-batch cultivation process. The cultivation was performed in the *Institute of Technical Chemistry, of the University of Hannover, Germany* during the collaboration work with the *Institute of Biophysics and Biomedical Engineering, BAS, Bulgaria*, and was funded by a grant *DFG*. The experimental data set includes records for the substrate feeding rate, concentration of biomass and substrate (glucose), and the cultivation time. In the nonlinear mathematical model considered here, the parameters that should be estimated are the maximum specific growth rate (μ_{max}), the saturation constant (k_S), and the yield coefficient ($Y_{S/X}$).

The parameter estimation is performed based upon the use of a modified Hausdorff Metric [39] and the most commonly used metric – the Least Square Regression. The Hausdorff Metrics are used in the geometric settings for measuring the distance between sets of points. They have been used extensively in areas such as computer vision, pattern recognition and computational chemistry [48,43,26,9]. The modified Hausdorff Distance is proposed to evaluate the mismatch between the experimental and the model predicted data. The results from both metrics are compared and analyzed.

The rest of the paper is organized as follows. The optimal parameters setting problem is formulated in Section 2. The ACO algorithm for the considered problem is defined in Section 3. The GA algorithm is described in Section 4. The numerical results and the discussion are presented in Section 5. Concluding remarks are introduced in Section 6.

2   Problem Formulation

The costs of developing mathematical models for the bioprocess improvement are often too high and the benefits too low. The main reason for this is related to the intrinsic complexity and non-linearity of biological systems. In general, mathematical descriptions of growth kinetics assume extensive simplifications. These models are often not accurate enough to correctly describe the underlying mechanisms. Another critical issue is related to the nature of the bioprocess models. Quite often, the parameters involved are not identifiable. Additionally, from the practical point of view, such identification would require data from specific experiments, which are themselves difficult to design and to realize. However, the estimation of model parameters with high parameter accuracy is essential for successful model development.

The real parameter optimization of simulation models, has become a research field of great interest in recent years. Nevertheless, after all completed research, this task still represents a very difficult problem. This mathematical problem, the so-called inverse problem, is a big challenge for the traditional optimization methods. In this case only the direct optimization strategies can be applied, because they exclusively use information about values of the goal function. Additional information about the goal function, like its gradients, etc., which could be used to accelerate the optimization process, is not available. Since an evolution of a goal for one string is provided by one simulation run, completing of the optimization algorithm may require a lot of computation time. Therefore, various metaheuristics are used as an alternative to surmount the parameter estimation difficulties.

2.1   Problem Model

The general state space dynamical model of the process of interest was described by Bastin and Dochain in [4]. It is accepted as representing the dynamics of an n components and m reactions bioprocess:

$$\frac{dx}{dt} = K\varphi(x,t) - Dx + F - Q. \tag{1}$$

Here, x is a vector representing the state components; K is the yield coefficient matrix; φ is the growth rates vector; the vectors F and Q are the feed rates and the gaseous outflow rates. The scalar D is the dilution rate, which will be the manipulated variable, and which is defined as follows:

$$D = \frac{F_{in}}{V} \tag{2}$$

where F_{in} is the influent flow rate and V is the bioreactor volume.

Application of the general state space dynamical model [4] to the *E. coli* cultivation fed-batch process leads to the following nonlinear differential equation system [33]:

$$\frac{dX}{dt} = \mu_{max}\frac{S}{k_S + S}X - \frac{F_{in}}{V}X \tag{3}$$

$$\frac{dS}{dt} = -\frac{1}{Y_{S/X}}\mu_{max}\frac{S}{k_S + S}X + \frac{F_{in}}{V}(S_{in} - S) \tag{4}$$

$$\frac{dV}{dt} = F_{in} \tag{5}$$

where:

X – biomass concentration, [g/l];
S – substrate concentration, [g/l];
F_{in} – feeding rate, [l/h];
V – bioreactor volume, [l];
S_{in} – substrate concentration in
          the feeding solution, [g/l];
μ_{max} – maximum value of
          the specific growth rate, $[h^{-1}]$;
k_S – saturation constant, [g/l];
$Y_{S/X}$ – yield coefficient, [-].

The mathematical formulation of the nonlinear dynamic model (Eqs. (3) - (5)) of the *E. coli* fed-batch cultivation process is described according to the mass balance and the model is based on the following a'priori assumptions:

− the bioreactor is completely mixed;
− the main products are biomass, water and, under some conditions, acetate;

- the substrate glucose is consumed mainly oxidatively and its consumption can be described by the Monod kinetics;
- the variation in the growth rate and the substrate consumption do not significantly change the elemental composition of the biomass, thus only balanced growth conditions are assumed;
- parameters, e.g. temperature, pH, or pO_2, are controlled at their individual constant set points.

For the parameter estimation problem the real experimental data of the *E. coli MC4110* fed-batch cultivation process is used. Off-line measurements of the biomass and on-line measurements of the glucose concentration are used in the identification procedure. The cultivation condition and the experimental data have been published in [34]. Here only the fermentation conditions described.

The fed-batch cultivation of the *E. coli* MC4110 is performed in a 2l bioreactor (Bioengineering, Switzerland), using a mineral medium [3], in the *Institute of Technical Chemistry, University of Hannover*. Before inoculation, a glucose concentration of 2.5 g/l is established in the medium. Glucose in the feeding solution is 100 g/l. The initial liquid volume is 1350 ml. The pH is controlled at 6.8 and the temperature is kept constant at 35°C. The aeration rate is kept at 275 l/h air, the stirrer speed at start 900 rpm, and after 11 hours the stirrer speed is increased in steps of 100 rpm. At end the stirrer sped reaches 1500 rpm. Oxygen is controlled at around 35%.

Off-line analysis
For the off-line glucose measurements, as well as the biomass and the acetate concentration determination, samples of about 10 ml are taken approximately at every hour. Off-line measurements are performed by using the Yellow Springs Analyser (Yellow Springs Instruments, USA).

On-line analysis
For the on-line glucose determination a flow injection analysis (FIA) system has been employed, using two pumps (ACCU FM40, SciLog, USA) for the continuous sample and the carrier flow rate. To reduce the measurement noise the continuous-discrete extended Kalman filter was used [3].

Glucose measurement and control system
For on-line glucose determination, the same FIA system has been employed for the continuous sample and the carrier flow rate at 0.5 ml/min and 1.7 ml/min respectively. A total of 24 ml of cells containing the culture broth were injected into the carrier stream and mixed with an enzyme solution of 350 000 U/l of glucose oxidase (Fluka, Germany) of a volume of 36 ml. After passing a reaction coil of 50 cm length, the oxygen uptake was measured using an oxygen electrode (ANASYSCON, Germany). To determine the oxygen consumed by cells only, no enzyme solution were injected. Calculating the difference of both

dissolved oxygen peak Heights, the glucose concentration can be determined. The time between sample taking and the measurement of the dissolved oxygen was $\Delta t = 45$ s.

For the automation of the FIA system, as well as glucose concentration determination, the software CAFCA (ANASYSCON, Germany) was applied. To reduce the measurement noise the continuous-discrete extended Kalman filter was used. This program was running on a separate PC and got the measurement results via a serial connection. A PI controller was applied to adjust the glucose concentration to the desired set point of 0.1 g/l [3].

The initial process conditions were [3]:

$t_0 = 6.68$ h, $X(t_0) = 1.25$ g/l, $S(t_0) = 0.8$ g/l, $S_{in} = 100$ g/l.

2.2 Optimization Criterion

From the practical perspective, modeling studies are performed to identify simple and easy-to-use models that are suitable to support the engineering tasks of process optimization and, especially, of control. The most appropriate model must satisfy the following conditions:

(i) the model structure should be able to represent the measured data in a proper manner;
(ii) the model structure should be as simple as possible, while remaining compatible with the first requirement.

On account of that, the cultivation process dynamic is described using a simple Monod-type model, the most common kinetics applied for modelling of the cultivation processes [4].

The optimization criterion is a certain factor, value of which defines the quality of an estimated set of parameters. To evaluate the mishmash between the experimental and the model predicted data, a modified Hausdorff Distance and the Least Square Regression are proposed.

In this work the Hausdorff Metric is used for the first time to solve the parameter optimization problem involving cultivation processes models.

Hausdorff Distance. When talking about distances, it usually means the shortest: for instance, if a point X is said to be at distance D of a polygon P, it is generally assumed that D is the distance from X to the nearest point of P. The same logic applies for polygons: if two polygons A and B are at some distance from each other, it commonly understood that the distance is the shortest one between any point of A and any point of B. That definition of distance between polygons can become quite unsatisfactory for some applications. However, it would be natural to expect that a small distance between two polygons means that no point of one polygon is far from the other polygon. Unfortunately, the shortest distance concept carries very low informative content.

In mathematics, the Hausdorff Distance, or the Hausdorff Metric (named after Felix Hausdorff), also called Pompeiu-Hausdorff Distance [39], measures how far two subsets of a metric space are from each other. It turns the set of non-empty compact subsets of a metric space into a metric space in its own right. Informally, two sets are close in the Hausdorff Distance if every point of either set is close to some point of the other set. In other words, the Hausdorff Distance is the longest distance you can be forced to travel by an adversary who chooses a point in one of the two sets, from where you then must travel to the other set. Thus, it is the farthest point of a set that you can be at, to the closest point of a different set. More formally, the Hausdorff Distance from set A to set B is a maxmin function defined as:

$$h(A, B) = \max_{a \in A} \left\{ \min_{b \in B} \{d(a, b)\} \right\},$$

(6)

where a and b are points of sets A and B respectively, and $d(a, b)$ is any metric between these points. For simplicity, in this work, the $d(a, b)$ as the Euclidean distance between a and b is taken. If sets A and B are made of lines or polygons instead of single points, then $h(A, B)$ applies to all defining points of these lines or polygons, and not only to their vertices. The Hausdorff Distance gives an interesting measure of mutual proximity, by indicating the maximal distance between any point of one set to the other set. IN this way it is better than the shortest distance, which applied only to one point of each set, irrespective of all other points of the sets.

Least Squares Regression. The objective of the modeling process consists of adjusting the parameters of a model function to best fit the data set. A simple data set consists of n points (data pairs) (x_i, y_i), $i = 1, 2, \ldots, n$, where x_i is an independent variable and y_i is a dependent variable value of which is found by observation. The model function has the form $f(x, \beta)$, where the m adjustable parameters are held in the vector β. The goal is to find the parameter values for the model which "best" fits the data. The least squares method finds its optimum when the sum S of squared residuals:

$$S = \sum_{i=1}^{n} r_i^2$$

is at a minimum. A residual is defined as the difference between the actual value of the dependent variable and the value predicted by the model. A data point may consist of more than one independent variable. For example, when fitting a plane to a set of height measurements, the plane is a function of two independent variables, x and z. In the most general case there may be one or more independent variables and one or more dependent variables at each data point.

$$r_i = y_i - f(x_i, \beta).$$

3   Ant Colony Optimization (ACO)

The ACO is a stochastic optimization method that mimics the social behavior of real ants colonies, which manage to establish the shortest rout to feeding sources and back. Real ants foraging for food lay down quantities of pheromone (chemical cues) marking the path that they follow. An isolated ant moves essentially at random but an ant encountering a previously laid pheromone will detect it and decide to follow it with high probability and thereby reinforce it with a further quantity of pheromone. The repetition of the above mechanism represents the auto-catalytic behavior of a real ant colony, where the more the ants follow a trail, the more attractive that trail becomes. The original idea comes from observing the exploitation of food resources among ants, in which ants' individually limited cognitive abilities have collectively been able to find the shortest path between a food source and the nest.

Basic of Ant Algorithm
The ACO is usually implemented as a team of intelligent agents, which simulate the ants behavior, walking around the graph representing the problem to solve, using mechanisms of cooperation and adaptation. The requirements of the ACO algorithm are as follows [6,13]:

- The problem needs to be represented appropriately, which would allow the ants to incrementally update the solutions through the use of a probabilistic transition rules, based on the amount of pheromone in the trail and other problem specific knowledge.
- A problem-dependent heuristic function, that measures the quality of components that can be added to the current partial solution.
- A rule set for pheromone updating, which specifies how to modify the pheromone value.
- A probabilistic transition rule based on the value of the heuristic function and the pheromone value, that is used to iteratively construct a solution.

The structure of the ACO algorithm is shown by the pseudocode below (Figure 1). The transition probability $p_{i,j}$, to choose the node j when the current node is i, is based on the heuristic information $\eta_{i,j}$ and the pheromone trail level $\tau_{i,j}$ of the move, where $i, j = 1, \ldots, n$.

$$p_{i,j} = \frac{\tau_{i,j}^a \eta_{i,j}^b}{\sum\limits_{k \in Unused} \tau_{i,k}^a \eta_{i,k}^b}, \tag{7}$$

where $Unused$ is the set of unused nodes of the graph.

The higher the value of the pheromone and the heuristic information, the more profitable it is to select this move and resume the search. In the beginning, the initial pheromone level is set to a small positive constant value τ_0; later, the ants update this value after completing the construction stage. The ACO algorithms adopt different criteria to update the pheromone level.

Ant Colony Optimization
Initialize number of ants;
Initialize the ACO parameters;
while not end-condition **do**
    for $k = 0$ **to** number of ants
        ant k choses start node;
        while solution is not constructed **do**
            ant k selects higher probability node;
        end while
    end for
    Update-pheromone-trails;
end while

Fig. 1. Pseudocode for ACO

The pheromone trail update rule is given by:

$$\tau_{i,j} \leftarrow \rho\tau_{i,j} + \Delta\tau_{i,j}, \tag{8}$$

where ρ models evaporation in the nature and $\Delta\tau_{i,j}$ is new added pheromone which is proportional to the quality of the solution. Thus better solutions will receive more pheromone than others and will be more desirable in a next iteration.

4   Genetic Algorithm

GA originated from the studies of cellular automata, conducted by John Holland and his colleagues at the University of Michigan. Holland's book [19], published in 1975, is generally acknowledged as the beginning of the research of genetic algorithms. The GA is a model of machine learning which derives its behavior from a metaphor of the processes of evolution in nature [18]. This is done by the creation within a machine of a population of individuals represented by chromosomes. A chromosome could be an array of real numbers, a binary string, a list of components in a database, all depending on the specific problem. The GA are highly relevant for industrial applications, because they are capable of handling problems with non-linear constraints, multiple objectives, and dynamic components – properties that frequently appear in the real-world problems [24,18]. Since their introduction and subsequent popularization [19], the GA have been frequently used as an alternative optimization tool to the conventional methods [18,29] and have been successfully applied in a variety of areas, and still find increasing acceptance [28,11,1,5,37,38,2].

Basic of Genetic Algorithm. GA was developed to model adaptation processes mainly operating on binary strings and using a recombination operator with mutation as a background operator. The GA maintains a population of individuals, $P(t) = x_1^t, ..., x_n^t$ for generation t. Each individual represents a potential solution to the problem and is implemented as some data structure S.

Each solution is evaluated to give some measure of its "fitness". Fitness of an individual is assigned proportionally to the value of the objective function of the individuals. Then, a new population (generation $t + 1$) is formed by selecting more fit individuals (selected step). Some members of the new population undergo transformations by means of "genetic" operators to form new solution. There are unary transformations m_i (mutation type), which create new individuals by a small change in a single individual ($m_i : S \rightarrow S$), and higher order transformations c_j (crossover type), which create new individuals by combining parts from several individuals ($c_j : S \times \ldots \times S \rightarrow S$). After some number of generations the algorithm converges - it is expected that the best individual represents a near-optimum (reasonable) solution. The combined effect of selection, crossover and mutation gives so-called reproductive scheme growth equation [18]:

$$\xi\left(S, t+1\right) \geq \xi\left(S, t\right) \cdot eval\left(S, t\right) / \bar{F}\left(t\right) \left[1 - p_c \cdot \frac{\delta\left(S\right)}{m - 1} - o\left(S\right) \cdot p_m\right].$$

Differences that separate genetic algorithms from the more conventional optimization techniques could be defined as follows [18]:

1. Direct manipulation of a coding – GA work with a coding of the parameter set, not the parameter themselves;
2. GA search in a population of points, not a single point;
3. GA use payoff (objective function) information, not derivatives or other auxiliary knowledge;
4. GA use probabilistic transition rules (stochastic operators), not deterministic rules.

Compared with traditional optimization methods, GA simultaneously evaluates many points in the parameter space. It is more probable to converge towards the global solution. A genetic algorithm does not assume that the space is differentiable or continuous and can also iterate many times on each data received. A GA requires only information concerning the quality of the solution produced by each parameter set (objective function value information). This characteristic differs from optimization methods that require derivative information or, worse yet, complete knowledge of the problem structure and parameters. Since GA do not demand such problem-specific information, they are more flexible than most search methods. Also GA do not require linearity in the parameters which is needed in iterative searching optimization techniques. Genetic algorithms can solve hard problems, are noise tolerant, easy to interface to existing simulation models, and easy to hybridize. Therefore, this property makes genetic algorithms suitable and more workable in use for a parameter estimation of considered here cultivation process models. Moreover, the GA effectiveness and robustness have been already demonstrated for identification of fed-batch cultivation processes [8,36,35,32].

The structure of the GA is shown by the pseudocode below (Figure 2).

The population at time t is represented by the time-dependent variable P, with the initial population of random estimates being $P(0)$. Here, each decision variable in the parameter set is encoded as a binary string (with precision

```
begin
    i = 0
    Initial population P(0)
    Evaluate P(0)
    while (not done) do (test for termination criterion)
    begin
        i = i + 1
        Select P(i) from P(i − 1)
        Recombine P(i)
        Mutate P(i)
        Evaluate P(i)
    end
end
```

Fig. 2. Pseudocode for GA

of binary representation). The initial population is generated using a random number generator that uniformly distributes numbers in the desired range. The objective function (see Eq. (15)) is used to provide a measure of how individuals have performed in the problem domain.

5 Numerical Results and Discussion

In this section, a more precise description, concerning the application of the ACO and GA for the parameter optimization of the *E. coli* cultivation process model, is presented. Here, the parameters μ_{max}, k_S and $Y_{S/X}$ have to be estimated. For applying ACO first, the problem is represented by a graph. It is needed to find the optimal values of three parameters which are interrelated. Therefore, the problem is represented with three-partitive graph. The graph consists of three levels. Every level represents a search area of one of the parameters that will be optimized. Every area is thus discretized, to consists of 1000 points (nodes), which are uniformly distributed in the search interval of every parameter. The first level of the graph represents the parameter μ_{max}. The second level represents the parameter k_S. The third level represents the parameter $Y_{S/X}$. There are arcs between nodes from consecutive levels of the graph and there are no arcs between nodes from the same level. The pheromone is deposited on the arcs, to indicate how good is this parameter combination. Every level of the graph of the problem consists of 1000 points, thus the number of possible solutions is 10^9, therefore is unpractical to apply the exact methods.

5.1 ACO for Parameter Optimization

Here the proposed ACO approach is very close to real ant behavior. Starting to create a solution, the ants chose a node from the firs level in a random way. Next, for nodes from the second and the third level, they apply the probabilistic rule.

The transition probability depends only on the pheromone level. The heuristic information is not used. Thus the transition probability is as follows:

$$p_{i,j} = \frac{\tau_{i,j}}{\sum\limits_{k \in Unused} \tau_{i,k}}, \tag{9}$$

The ants prefer the node with maximal probability, which is the node with maximal quantity of the pheromone on the arc (starting from the current node). If there is more than one candidate for next node, the ant chooses randomly between the candidates. The process is iterative. At the end of every iteration the pheromone on the arcs is updated. The quality of the solutions is represented by the value of the objective function. In this case the objective function is the mean distance between the simulated data and the experimental data, which are the concentration of the biomass and the concentration of the substrate. The aim of the process is to minimize it, therefore the new added pheromone by ant i is:

$$\Delta\tau = (1 - \rho)/J(i) \tag{10}$$

where $J(i)$ is the value of the objective function according the solution constructed by ant i. Thus the arcs corresponding to solutions with the lesser value of the objective function will receive more pheromone and will be more desirable in the next iteration.

The values of the parameters of the ACO algorithms are very important, because they manage the search process. Therefore, it is necessary to find appropriate parameter settings, where the number of ants is the main parameter. In the ACO a small number of ants between 10 and 20 can be used, without need to increase the number of iterations to achieve good solutions. The next parameter is the initial pheromone. Normally it has a small value. The last parameter is the evaporation rate, which shows the importance of the last found solution, as related to the previous ones. Parameters of the ACO were tuned based on several pre-tests according considered here optimization problem. After tuning procedures the main algorithm parameters are set to the optimal settings. The parameter settings for the ACO are shown in Table 1.

Table 1. Parameters of ACO algorithm

Parameter	Value
number of ants	20
initial pheromone	0.5
evaporation	0.1

5.2   GA for Parameter Optimization

The strings of artificial genetic systems are analogous to chromosomes in biological systems. Thus a chromosome representation is needed to describe each individual in the population. The representation scheme determines the genetic operators that are used. Each individual or chromosome is made up of a sequence of genes from a certain alphabet. Here applied alphabet consists of binary digits 0 and 1. A binary 20 bit representation is here considered.

A common selection approach assigns a probability of selection, P_j, to each individual j. A series of N random numbers is generated and compared against the cumulative probability, $C_i = \sum_{j=1}^{i} P_j$ of the population. The appropriate individual, i, is selected and copied into the new population if $C_{i-1} < U(0,1) \leq C_i$. Roulette wheel, developed by Holland [19] is the first selection method. The probability, P_i, for each individual is defined by:

$$P[\text{ Individual } i \text{ is chosen}] = \frac{F_i}{\sum_{j=1}^{PopSize} F_j}, \tag{11}$$

where F_i equals the fitness of individual i and $PopSize$ is the population size.

The genetic operators provide the basic search mechanism of the GA. The operators are used to create new solutions based on existing solutions in the population. There are two basic types of operators: crossover and mutation. The crossover takes two individuals and produces two new individuals. The mutation alters one individual to produce a single new solution. Let $\overline{X}$ and $\overline{Y}$ be two m-dimensional row vectors denoting individuals (parents) from the population. For $\overline{X}$ and $\overline{Y}$ binary, the following operators are defined: binary mutation and simple crossover.

Binary mutation flips each bit in every individual in the population with probability p_m according to Eq. (12) [20]:

$$x_i = \begin{cases} 1 - x_i, & \text{if } U(0,1) < p_m \\ x_i, & \text{otherwise} \end{cases}. \tag{12}$$

Simple crossover generates a random number r from a uniform distribution from 1 to m and creates two new individuals $\overline{X'}$ and $\overline{Y'}$ according to Eqs. (13) and (13) [20].

$$x_i' = \begin{cases} x_i, & \text{if } i < r \\ y_i, & \text{otherwise} \end{cases}. \tag{13}$$

$$y_i' = \begin{cases} y_i, & \text{if } i < r \\ x_i, & \text{otherwise} \end{cases}. \tag{14}$$

In proposed genetic algorithm fitness-based reinsertion (selection of offspring) is used. The GA moves from generation to generation selecting and reproducing parents until a termination criterion is met. The most frequently used stopping criterion is a specified maximum number of generations. Values of genetic algorithm parameters are listed in Table 2.

In this paper two different measures – the Least Square Regression and the modified Hausdorff Distance are compared. The modified Hausdorff Distance,

Table 2. Parameters of GA

Parameter	Value
ggap	0.97
xovr	0.75
mutr	0.01
nind	100
maxgen	200

which is conformable to the considered problem is applied. There are two sets of points, the simulated (model predicted) and the measured (experimental) data, which form two lines. The Euclidean distance $d(t)$ between points from the two lines, corresponding to the same time moment t, is calculated. After that, the Euclidean distance from the point of one of the lines in time t to the points from other line in the time interval $(t - d(t), t + d(t))$ is calculated, and the minimal of these distances is taken. This is the distance between the two lines in the time moment t. Thus, the number of calculations, compared with the traditional Hausdorff Distance, is decreased, due to the fact that the distance to the points out of the interval $(t - d(t), t + d(t))$ will be large. At the end all distances between the points and the lines are combined. Thereby, eventual larger distance in some time moment, due to the measurement noise, is eliminated.

Thus, the objective function is presented as a minimization of the modified Hausdorff Distance measure J_1 between experimental and model predicted values of the state variables, represented by the vector **y**:

$$J_1 = \sum_{i=1}^{m} h\left(\mathbf{y}_{\exp}(i), \mathbf{y}_{\bmod}(i)\right)^2 \to \min \tag{15}$$

where m is the number of state variables (biomass and glucose concentrations); $\mathbf{y}_{\exp}$ is the known experimental data; while $\mathbf{y}_{\bmod}$ model predictions with a given set of the parameters.

In the case of the Least Square Regression the objective function is:

$$J_2 = \sum_{i=1}^{m} \left(\mathbf{y}_{\exp}(i) - \mathbf{y}_{\bmod}(i)\right)^2 \to \min \tag{16}$$

5.3   Numerical Results

All experiments have been conduced on a PC with Intel Core 2 2.8 GHz, 3.5 GB Memory, Linux operating system and using the Matlab 7.5 environment.

Because of the stochastic characteristics of the applied ACO and GA algorithms, a series of 30 runs for each algorithm was performed.

To study the algorithm performance, the worst the best and the average results of the 30 runs, for the objective function values of the two variants of ACO and GA algorithms are studied. For a realistic comparison, the number of iterations is

Table 3. ACO and GA with Least Square Regression and Hausdorff Distance

	Method	Average	Worst	Best
ACO Least Square	–	4.8866	6.7700	3.3280
	Hausdorff	2.3875	4.1290	1.7218
ACO Hausdorff	–	1.8744	2.5322	1.6425
	Least Square	3.9706	4.4283	3.4276
GA Least Square	–	4.8341	5.4234	4.7314
	Hausdorff	1.7510	2.0224	1.7025
GA Hausdorff	–	2.0299	2.3326	1.7657
	Least Square	4.3549	4.8872	3.5464

fixed to be 100. The average, worst and best values of the objective functions are shown on Table 3. In the first line of the second row, the average, worst and best value of the objective function are shown when Least Square Regression is used with ACO. The second line in the second row depicts the calculated Hausdorff Distance between the same solutions achieved when the objective function is the Least Square Regression. The first line of the third row shows the average, best and worst values of the objective function when it is the Hausdorff Distance with ACO. The second line of the third row represents the Least Square Distance of the same solutions achieved when the objective function is the Hausdorff Distance. Comparing the two rows it can be observed that the average and the worst achieved results are much better using the Hausdorff Distance than the Least Square Regression. The best achieved solutions are similar. In the best achieved solutions it can be seen that the Hausdorff Distance between achieved solutions is smaller when the objective function is Hausdorff, but the Least Square Distance is smaller when the objective function is the Least Squares Regression. During a number of runs of the algorithm the same phenomenon was observed – a small Hausdorff Distance between modeled and measured data and at the same time a big Least Square Distance between the data. When the Least Squares Regression is applied as the metric, the distance between the two lines can be very big, and in the same time it is seen that they are geometrically close to each other. It can happen especially in the steep parts of the lines. Applying the Hausdorff Metrics it can not happen, because it measures the geometrical similarity. Overall, the Hausdorff Distance is more time consuming than Least Square Distance, but much more realistic for the type of problems considered here. It can be concluded that ACO algorithm proposed in this paper performs better when the objective function is the Hausdorff Distance. We do the similar analysis when we apply GA. In this case there is very small difference between achieved results when we apply Hausdorff distance and Least Square Regression. We can conclude that GA is less sensitive according used measures.

In Table 4 the best parameter values (μ_{max}, k_S and $Y_{S/X}$), obtained using the ACO with the objective function based on the Hausdorff Distance, are presented.

The obtained model dynamics compared to the real experimental data is presented in Fig. 2 and Fig. 3.

Table 4. Best parameter values of the model

Parameter	Value
μ_{max}	0.5283
k_S	0.0174
$Y_{S/X}$	2.0300

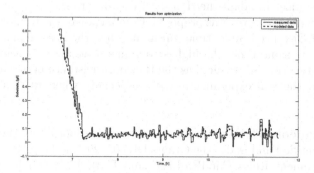

Fig. 3. Time profiles of the substrate: experimental data and models predicted data

In Figure 3, the modelled substrate is represented by the dash line, while by solid line the measured substrate is depicted. In Figure 4, line represents values of the modelled biomass, while stars represent values of the measured biomass.

The presented figures show a very good correlation between the experimental and model predicted data and confirm the obtained results.

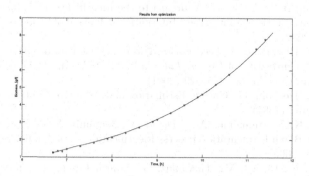

Fig. 4. Time profiles of the biomass: experimental data and models predicted data

6   Concluding Remarks

In this work the ACO and GA algorithms for a parameter setting of the *E. coli* fed-batch cultivation process model ware proposed. The methods are chosen as the most common direct methods used for global optimization. The process model is presented as a system of nonlinear ordinary differential equation describing the biomass and the substrate dynamics. In the identification procedures, the real experimental data was used. The objective function was formulated as the difference between the modeled and the measured data. When solving the optimization problem, two different measures were used – the commonly used Least Square Regression and, for the first time applied to this type of problem, the Hausdorff Distance. To adapt the Hausdorff Distance to the considered problem a modification of this metric was proposed. Comparison of the results shows that the Hausdorff Distance is more time consuming than the Least Square Distance. However, at the same time, the highest parameter accuracy is achieved when the objective function is measured as the Hausdorff Distance between the model predicted and the real experimental data, especially when ACO algorithm is applied.

Acknowledgment. This work has been partially supported by the Bulgarian National Scientific Fund under the Grants DID 02/29 "Modeling Processes with Fixed Development Rules (ModProFix)" and DMU 02/4 "High quality control of biotechnological processes with application of modified conventional and metaheuristics methods". Work presented here is a part of the Poland-Bulgarian collaborative Grant "Parallel and distributed computing practices".

References

1. Akpinar, S., Bayhan, G.M.: A Hybrid Genetic Aalgorithm for Mixed Model Assembly Line Balancing Problem with Parallel Workstations and Zoning Constraints. Engineering Applications of Artificial Intelligence 24(3), 449–457 (2011)
2. Al-Duwaish, H.N.: A Genetic Approach to the Identification of Linear Dynamical Systems with Static Nonlinearities. International Journal of Systems Science 31(3), 307–313 (2000)
3. Arndt, M., Hitzmann, B.: Feed Forward/feedback Control of Glucose Concentration during Cultivation of *Escherichia coli*. In: 8th IFAC Int. Conf. on Comp. Appl. in Biotechn, Canada, pp. 425–429 (2001)
4. Bastin, G., Dochain, D.: On-line Estimation and Adaptive Control of Bioreactors Els. Sc. Publ. (1991)
5. Benjamin, K.K., Ammanuel, A.N., David, A., Benjamin, Y.K.: Genetic Algorithm using for a Batch Fermentation Process Identification. J. of Applied Sciences 8(12), 2272–2278 (2008)
6. Bonabeau, E., Dorigo, M., Theraulaz, G.: Swarm Intelligence: From Natural to Artificial Systems. Oxford University Press, New York (1999)
7. Brownlee J.,: Clever Algorithms. Nature-Inspired Programming Recipes, LuLu, p. 436, 978-1-4467-8506-5 (2011)
8. Carrillo-Ureta, G.E., Roberts, P.D., Becerra, V.M.: Genetic Algorithms for Optimal Control of Beer Fermentation. In: Proc. of the 2001 IEEE International Symposium on Intelligent Control, Mexico City, Mexico, pp. 391–396 (2001)

9. Chen, S., Lovell, B.C.: Feature space Hausdorff distance for face recognition. In: Proc. of 20th International Conference on Pattern Recognition (ICPR), Istanbul, Turkey, pp. 1465–1468 (2010)

10. Covert, M.W., Xiao, N., Chen, T.J., Karr, J.R.: Integrating Metabolic, Transcriptional Regulatory, and Signal Transduction Models in *Escherichia coli*. J. of Bioinformatics 24(18), 2044–2050 (2008)

11. da Silva, M.F.J., Perez, J.M.S., Pulido, J.A.G., Rodriguez, M.A.V.: AlineaGA - A Genetic Algorithm with Local Search Optimization for Multiple Sequence Alignment. Appl. Intell. 32, 164–172 (2010)

12. Dorigo, M., Di Caro, G.: The Ant Colony Optimization Meta-heuristic. In: Corne, D., Dorigo, M., Glover, F. (eds.) New Idea in Optimization, pp. 11–32. McGrow-Hill (1999)

13. Dorigo, M., Stutzle, S.: Ant Colony Optimization. MIT Press (2004)

14. Fidanova, S.: ACO algorithm with additional reinforcement. In: Dorigo, M., Di Caro, G.A., Sampels, M. (eds.) ANTS 2002. LNCS, vol. 2463, pp. 292–293. Springer, Heidelberg (2002)

15. Fidanova, S., Lirkov, I.: 3D Protein Structure Prediction. J. Analele Universitatii de Vest Timisoara, Seria Matematica-Informatica XLVII(2), 33–46 (2009) ISSN 1224-970X

16. Fidanova, S.: An Improvement of the Grid-based Hydrophobic-hydrophilic Model. Int. J. Bioautomation 14(2), 147–156 (2010) ISSN 1312-451X

17. Fidanova, S., Alba, E., Molina, G.: Hybrid ACO algorithm for the GPS surveying problem. In: Lirkov, I., Margenov, S., Waśniewski, J. (eds.) LSSC 2009. LNCS, vol. 5910, pp. 318–325. Springer, Heidelberg (2010)

18. Goldberg, D.E.: Genetic Algorithms in Search, Optimization and Machine Learning. Addison Wesley Longman, London (2006)

19. Holland, J.H.: Adaptation in Natural and Artificial Systems, 2nd edn. MIT Press, Cambridge (1992)

20. Houck, C.R., Joines, J.A., Kay, M.G.: A Genetic Algorithm for Function Optimization: A Matlab Implementation. Genetic Algorithm Toolbox Toutorial (1996), http://read.pudn.com/downloads152/ebook/662702/gaotv5.pdf

21. Jiang, L., Ouyang, Q., Tu, Y.: Quantitative Modeling of *Escherichia coli* Chemotactic Motion in Environments Varying in Space and Time. PLoS Comput. Biol. 6(4), e1000735 (2010), doi:10.1371/journal.pcbi.1000735

22. Karelina, T.A., Ma, H., Goryanin, I., Demin, O.V.: EI of the Phosphotransferase System of *Escherichia coli*: Mathematical Modeling Approach to Analysis of Its Kinetic Properties. Journal of Biophysics 2011, Article ID 579402 (2011), doi:10.1155/2011/579402

23. Kirkpatrick, S., Gelatt, C.D., Vecchi, M.P.: Optimization by Simulated Annealing. Science, New Series 220(4598), 671–680 (1983)

24. Kumar, S.M., Giriraj, R., Jain, N., Anantharaman, V., Dharmalingam, K.M.M., Sheriffa, B.: Genetic algorithm based PID controller tuning for a model bioreactor. Indian Chemical Engineer 50(3), 214–226 (2008)

25. Michalewicz, Z.: Genetic Algorithms + Data Structures = Evolution Programs, 2nd Exended edn. Springer, Heidelberg (1994)

26. Nutanong, S., Jacox, E.H., Samet, H.: An Incremental Hausdorff Distance Calculation Algorithm. Proc. of the VLDB Endowment 4(8), 506–517 (2011)

27. Opalka, N., Brown, J., Lane, W.J., Twist, K.-A.F., Landick, R., Asturias, F.J., Darst, S.A.: Complete Structural Model of *Escherichia coli* RNA Polymerase from a Hybrid Approach. PLoS Biol. 8(9), e1000483 (2010), doi:10.1371/journal.pbio.1000483

28. Paplinski, J.P.: The Genetic Algorithm with Simplex Crossover for Identification of Time Delays. Intelligent Information Systems, 337–346 (2010)
29. Parker, B.S.: Demonstration of using Genetic Algorithm Learning. Information Systems Teaching Laboratory (1992)
30. Pardalos, P.M., Resende, M.G.C.: Handbook of Applied Optimization. Oxford University Press (2002)
31. Petersen, C.M., Rifai, H.S., Villarreal, G.C., Stein, R.: Modeling *Escherichia coli* and Its Sources in an Urban Bayou with Hydrologic Simulation Program – FORTRAN. Journal of Environmental Engineering 137(6), 487–503 (2011)
32. Ranganath, M., Renganathan, S., Gokulnath, C.: Identification of Bioprocesses using Genetic Algorithm. Bioprocess Engineering 21, 123–127 (1999)
33. Roeva, O.: Parameter estimation of a monod-type model based on genetic algorithms and sensitivity analysis. In: Lirkov, I., Margenov, S., Waśniewski, J. (eds.) LSSC 2007. LNCS, vol. 4818, pp. 601–608. Springer, Heidelberg (2008)
34. Roeva, O., Pencheva, T., Hitzmann, B., Tzonkov, S.: A Genetic Algorithms Based Approach for Identification of *Escherichia coli* Fed-batch Fermentation. Int. J. Bioautomation 1, 30–41 (2004)
35. Roeva, O.: A Modified Genetic Algorithm for a Parameter Identification of Fermentation Processes. Biotechnology and Biotechnological Equipment 20(1), 202–209 (2006)
36. Roeva, O.: Multipopulation genetic algorithms: A tool for parameter optimization of cultivation processes models. In: Boyanov, T., Dimova, S., Georgiev, K., Nikolov, G. (eds.) NMA 2006. LNCS, vol. 4310, pp. 255–262. Springer, Heidelberg (2007)
37. Roeva, O.: Improvement of Genetic Algorithm Performance for Identification of Cultivation Process Models. In: Advances Topics on Evolutionary Computing, Book Series: Artificial Intelligence Series-WSEAS, pp. 34–39 (2008)
38. Roeva, O., Slavov, T.: Fed-batch cultivation control based on genetic algorithm PID controller tuning. In: Dimov, I., Dimova, S., Kolkovska, N. (eds.) NMA 2010. LNCS, vol. 6046, pp. 289–296. Springer, Heidelberg (2011)
39. Rote, G.: Computing the minimum Hausdorff distance between two point sets on a line under translation. Information Processing Letters 38, 123–127 (1991)
40. Schuegerl, K., Bellgardt, K.-H.: Bioreaction Engineering: Modeling and Control. Springer, Heidelberg (2000)
41. Shmygelska, A., Hoos, H.H.: An ant colony optimization algorithm for the 2D and 3D hydrophobic polar protein folding problem. BMC Bioinformatics 6(30) (2005), doi:10.1186/1471-2105-6-30
42. Skandamis, P.N., Nychas, G.E.: Development and Evaluation of a Model Predicting the Survival of *Escherichia coli* O157:H7 NCTC 12900 in Homemade Eggplant Salad at Various Temperatures, pHs, and Oregano Essential Oil Concentrations. Applied and Environmental Microbiology 66(4), 1646–1653 (2000)
43. Sugiyama, M., Hirowatari, E., Tsuiki, H., Yamamoto, A.: Learning figures with the hausdorff metric by fractals. In: Hutter, M., Stephan, F., Vovk, V., Zeugmann, T. (eds.) ALT 2010. LNCS, vol. 6331, pp. 315–329. Springer, Heidelberg (2010)
44. Syam, W.P., Al-Harkan, I.M.: Comparison of Three Meta Heuristics to Optimize Hybrid Flow Shop Scheduling Problem with Parallel Machines. World Academy of Science, Engineering and Technology 62, 271–278 (2010)
45. Tahouni, N., Smith, R., Panjeshahi, M.H.: Comparison of Stochastic Methods with Respect to Performance and Reliability of Low-temperature Gas Separation Processes. The Canadian Journal of Chemical Engineering 88(2), 256–267 (2010)

46. Umarani, R., Selvi, V.: Particle Swarm Optimization: Evolution, Overview and Applications. Int J of Engineering Science and Technology 2(7), 2802–2806 (2010)
47. Viesturs, U., Karklina, D., Ciprovica, I.: Bioprocess and Bioengineering, Jeglava (2004)
48. Yedjour, H., Meftah, B., Yedjour, D., Benyettou, A.: Combining Spiking Neural Network with Hausdorff Distance Matching for Object Tracking. Asian Journal of Applied Sciences 4, 63–71 (2011)
49. Yusof, M.K., Stapa, M.A.: Achieving of Tabu Search Algorithm for Scheduling Technique in Grid Computing using GridSim Simulation Tool: Multiple Jobs on Limited Resource. Int J of Grid and Distributed Computing 3(4), 19–31 (2010)

A Heuristic Based Algorithm for the 2D Circular Strip Packing Problem

Hakim Akeb[1], Mhand Hifi[2], and Dominique Lazure[2]

[1] ISC Paris School of Management
22 Boulevard du Fort de Vaux, 75017 Paris, France
[2] Université de Picardie Jules Verne, UR EPROAD, Équipe ROAD
7 rue du Moulin Neuf, 80039 Amiens, France

Abstract. This paper solves the strip packing problem (SPP) that consists in packing a set of circular objects into a rectangle of fixed width and unlimited length. The objective is to minimize the length of the rectangle that will contain all the objects such that no object overlaps another one. The proposed algorithm uses a look-ahead method combined with beam search and a restarting strategy. The particularity of this algorithm is that it can achieve good results quickly (faster than other known methods and algorithms) even when the number of objects is large. The results obtained on well-known benchmark instances from the literature show that the algorithm improves a lot of best known solutions.

Keywords: Cutting and packing, 2D strip packing, beam search, heuristic.

1 Introduction

Cutting & Packing (C&P) problems are well known in Operations Research since they have many practical applications. They are for example encountered in the storage and transportation of objects of different shapes (Baltacioglu *et al.* [1]; Bortfeldt and Homberger[2]; Castillo *et al.* [3]; Conway and Sloane [4]; Lewis *et al.* [5]). In this case, the objective is to arrange these objects in order to save space. C&P problems are also used in the industry when a set of pieces of predetermined shapes have to be cut from a rectangular plate (Menon and Schrage [6]). The objective in this second example is to minimize the waste due to the space between the pieces to cut.

This paper studies the problem of cutting (or packing) a set $N = \{1, .., n\}$ of n circular pieces C_i of known radii $r_i, i \in N$, from (or into) a strip S of fixed width W and unlimited length L. The objective is to place the n pieces inside the smallest rectangle R of dimensions $W \times L^*$ such that no piece overlaps another one and no piece exceeds the limits of the rectangle. This problem is known as the *Strip Packing Problem* or SPP (see Wäscher *et al.* [7]). Fig. 1 shows an instance containing nine circles ($N = \{1, .., 9\}$) to try to pack inside a rectangle of dimensions $W \times L$. Fig. 2 shows two feasible solutions in which all the circles

S. Fidanova (Ed.): *Recent Advances in Computational Optimization*, SCI 470, pp. 73–92.
DOI: 10.1007/978-3-319-00410-5_5   © Springer International Publishing Switzerland 2013

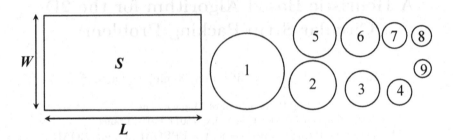

Fig. 1. Example of a set of nine circles to pack inside a rectangle $W \times L$

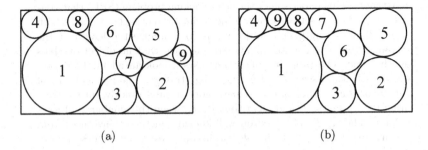

(a)   (b)

Fig. 2. Two feasible packings of the nine circles

were packed into the rectangle. Note that a rearrangement of the circles inside the rectangle may decrease its length. Indeed the solution displayed in Fig. 2 (b) is a rearrangement of the circles of the solution indicated in Fig. 2 (a). Note also that the most-right circle (5) in Fig. 2 (b) does not touch the right border of the rectangle, this means that the width of the rectangle can be decreased. This means also that this second solution (Fig. 2 (b)) is better than the one indicated in Fig. 2 (a) since the width of the rectangle is smaller. But a good rearrangement is not easy to be achieved because of the continuous characteristic of the variables (see below).

The mathematical formulation for SPP is as follows:

$$\min L \tag{1}$$

$$\sqrt{(x_i - x_j)^2 + (y_i - y_j)^2} \geq r_i + r_j, \text{ for } j < i, \ (i,j) \in N^2 \tag{2}$$

$$r_i \leq x_i \leq L - r_i, \ \forall i \in N, \tag{3}$$

$$r_i \leq y_i \leq W - r_i, \ \forall i \in N, \tag{4}$$

$$L \geq \frac{\pi}{W} \times \sum_{i=1}^{n} r_i^2 \tag{5}$$

Equation 1 indicates the objective to minimize, i.e., the length of the target rectangle that will contain the n pieces. Equation 2 means that any pair of distinct circles C_i and C_j do not overlap each other, i.e., the euclidean distance between their centers must be greater than or equal to the sum of their radii $r_i + r_j$. Equations 3–4 mean that any circle C_i does not exceed the container boundary. Finally, Equation 5 indicates that the objective to minimize (L) has a lower bound value, denoted by $\underline{L}$, which is equal the sum of the surfaces of the n circles divided by the width of the rectangle W. Any value for L cannot then be smaller than this lower bound otherwise this will mean that there is no space between the circles and between the circles and the container boudary.

A solution for the strip packing problem consists to find the minimum value for the length of the rectangle that will contain all the pieces while verifying the constraints represented by Equations 2–4.

2   Literature Review

The problem of packing circular objects of different radii into a container is well known and very studied in the literature. Since there is no method calculating exact solutions, the authors use generally heuristic-based approaches in order to compute approximate solutions for the problem. Two main categories of containers can be distinguished: the first one corresponds to a circle and the second one to a rectangle. In addition, the circular pieces may have the same radius or have different radii.

Packing different-sized circles into the smallest circle was for example studied by Huang et al. [8] where the authors proposed greedy algorithms based on the Maximum Hole Degree (MHD) heuristic. Hifi and M'Hallah [9] proposed a dynamic adaptive local search where the radius of the containing circle is increased when placing the circles. For the same problem, Akeb et al. [10] used beam-search based algorithms. Packing equal circles inside a circle was studied by several authors. Graham et al. [11] proposed two methods in order to pack a set of congruent circles inside the unit circle. The first method is called *Billiards simulation* and the second one is based of the *Energy Function Minimization*. Liu et al. [12] proposed a heuristic based on a technique called *Energy landscape paving* in order to pack equal circles inside the smallest containing circle.

The problem of packing circles of different radii into a rectangular container is more studied in the literature because of its various applications. For example George et al. [13] proposed several rules based essentially on the use of a genetic algorithm as well as a random strategy. Stoyan and Yaskov [14] designed a mathematical model whose objective is to search for feasible local optima by combining a tree-search procedure and a reduced gradient. A genetic algorithm was also used by Hifi and M'Hallah [15]. Huang et al. [16] designed two greedy algorithms for the strip packing problem, the algorithms, denoted by B1.0 and B1.5, are based on the Maximum Hole Degree (MHD) heuristic. Birgin et al. [17] used a non-linear approach for placing circles inside a rectangle. Kubach et al. [18] proposed a parallel version for the MHD heuristic for tackling the strip packing problem. Finally, Akeb *et al.* [19] proposed a beam-search based algorithm coupled with a restarting strategy for solving SPP.

Packing equal circles inside a square and/or a rectangle was for example studied by Huang and Ye [20], the authors proposed a stochastic method in order to place up to 200 circles. E. Specht [21] proposes a deterministic method in order to compute high density packings of equal circles in a rectangle. Several years before, Locatelli and Raber [22] used a branch-and-bound algorithm in order to pack a given number of equal circles into the unit square.

Some authors proposed several methods in order to place circles inside containers of different shapes. This is for example the case of López and Beasley [23] who used a non-linear formulation, involving Cartesian and polar coordinates, solved by the SNOPT solver. Birgin and Sobral [24] proposed several results for packing circles and spheres inside 2D and 3D containers. Finally, Birgin and Gentil [25] considered the packing of unitary radius circles inside triangles, rectangles, and strips.

In this paper, an improved algorithm is proposed for the strip packing problem. This algorithm combines beam search, a restarting strategy, and a look-ahead method. The objective of the look-ahead is to accelerate the search to obtain quickly solutions. In addition, the parameters of the restarting and the look-ahead strategies are studied in order to adapt them to the characteristics of each instance.

The rest of the paper is organized as follows. Section 3 explains how to use beam search in order to resolve the strip packing problem (SPP). Section 4 returns on some existing beam-search based algorithms for SPP. Section 5 details the improved algorithm denoted by IA. Section 6 discusses the results obtained by IA on the most known instances in the literature. Finally, Section 7 summarizes the results obtained and indicates some orientations for future work.

3 Beam Search for Resolving SPP

Beam search (BS) [26] is a tree-based search and is an adaptation of the best first search. BS selects, at each level ℓ of the tree, the most promising nodes to expand in order to create the nodes of the next level $\ell+1$. So a criterion, allowing to evaluate each node, must be defined. The number of the nodes chosen at each level is denoted by ω and is called the *beam width*.

A standard implementation of the BS method is given in Algorithm 1. BS receives two parameters: the root node B_0 that contains a starting solution (partial solution) and the value of the beam width ω. The algorithm's output is a feasible solution if it succeeds, or the empty set if not.

At line 1 of Algorithm 1, the root node is assigned to B (the set of nodes of the current level). The set of offspring nodes, i.e., the descendants of the nodes in B, is denoted by B_ω. After that, at line 2, the value of the best solution z^* is initialized to the best known solution if this one exists, otherwise z^* is set to $+\infty$, meaning that the problem at hand is a minimization (for a maximization, z^* is set to $-\infty$).

At each level of the search tree, i.e, the **while** loop (Lines 3–14), each node $\eta \in B$ generates several descendant nodes. Theses ones are inserted into the set B_ω (line 4). If a node in B_ω is a leaf (no branching is possible from it), then its solution value is computed (line 6) and the best solution z^* is updated if a better one is found. After that, the node is removed from B_ω (line 10). The other nodes in B_ω are after that evaluated by calculating their solution values and only the best ω nodes are kept, the other nodes are removed (line 12). The nodes chosen are then assigned to the set B and B_ω is reset to the empty set (line 13). The instructions in lines 3–14 are repeated until no branching is possible, i.e., $B = \emptyset$. At line 15, the algorithm returns the best solution found so far.

Note that the method described above is a width-first implementation of beam-search. Of course there also exists a depth-first implementation where the exploration goes as far as possible along each branch before backtracking. For more details, see [27].

Require: The root node B_0 (starting solution) and the beam width value ω.
Ensure: A feasible solution if such one is reached, the empty set otherwise.

1: Let $B = B_0$ be the set of nodes of the current level and B_ω the set of offspring nodes;
2: If a feasible solution is known then set z^* to its value, otherwise set $z^* = +\infty$;
3: **while** $(B \neq \emptyset)$ **do**
4:    Branch out of each node $\eta \in B$ and insert the resulting (offspring) nodes into B_ω;
5:    **if** a node $\eta_i \in B_\omega$ is a leaf **then**
6:       compute z_{η_i} the value of node η_i;
7:       **if** $z_{\eta_i} < z^*$ **then**
8:          update the best solution z^* $(z^* := z_{\eta_i})$;
9:       **end if**
10:      Remove z_{η_i} from B_ω;
11:   **end if**
12:   Keep only the ω best nodes in B_ω (those having the best values of z) and remove the others;
13:    $B := B_\omega$ and $B_\omega := \emptyset$;
14: **end while**
15: **return** the best solution if it exists, otherwise the empty set;

Algorithm 1. The Beam Search Method

The rest of this section is organized as follows. First, the different notations used throughout the paper are given. After that, a greedy procedure, denoted by MLDP (Minimum Local Distance Position) is described. The objective of MLDP is to try to place the n circles inside the current rectangle $R = W \times L$, i.e., when the length of the rectangle is fixed to a given value L.

3.1   Notations

In order to simplify the reading of the paper, here are the different notations used throughout the document:

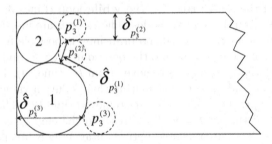

Fig. 3. The MLDP strategy

1. $N = \{1, ..., n\}$ is the set of circles to pack into the strip S placed with its bottom left corner at point $(0, 0)$ in the Euclidean plan,
2. $M = \{1, ..., m\}$ is the set of circle types (the set of different radii in the instance),
3. S_{left}, S_{top}, S_{right}, and S_{bottom} are the four edges of S,
4. The circular piece C_i of radius r_i is placed with its center at coordinates (x_i, y_i),
5. I_i corresponds to the set of circles already packed inside the strip ($|I_i| = i$),
6. $\bar{I}_i$ contains the circles not yet placed ($I_i \cup \bar{I}_i = N$),
7. P_{I_i} is the set of distinct corner positions for the next circle to place C_{i+1} given the set I_i,
8. A corner position $p_{i+1} \in P_{I_i}$ for C_{i+1} is computed by using two elements e_1 and e_2. An element is either a piece already placed (set I_i) or one of the three edges of S (S_{left}, S_{top}, S_{bottom}). $T_{p_{i+1}}$ denotes the set composed of both elements e_1 and e_2.

3.2   The MDLP Greedy Procedure

The Minimum Local Distance Position (MLDP) procedure can be used as a greedy algorithm in order to compute a solution. Indeed, given the set I_i of circles already placed inside the current rectangle and the set of corner positions P_{i+1} for the next circle C_{i+1}, MLDP selects the *best corner position* for this circle. This process is repeated until all the circles are placed or no additional circle can be placed. Fig 3 explains the mechanism of MLDP where two circles are already placed, thus $i = 2$ and $I_2 = \{C_1, C_2\}$. There are also three possible positions to place the next circle C_3: $P_{I_2} = \{p_3^{(k)}, k = 1, .., 3\}$. The first corner position $p_3^{(1)}$ touches circle C_2 and the top-edge of the strip S_{top}, then $T_{p_3^{(1)}} = \{C_2, S_{\text{top}}\}$. For the two others corner positions, $T_{p_3^{(2)}} = \{C_1, C_2\}$ and $T_{p_3^{(3)}} = \{C_1, S_{\text{bottom}}\}$.

Let C_{i+1} be the circular piece to place at position p_{i+1} and $\delta_{i+1}(edge)$, $edge \in E_{\text{edge}} = \{S_{\text{left}}, S_{\text{bottom}}, S_{\text{top}}\}$, the three distances defined as follows: $\delta_{i+1}(S_{\text{left}}) = x_{i+1} - r_{i+1}$, $\delta_{i+1}(S_{\text{bottom}}) = y_{i+1} - r_{i+1}$, and $\delta_{i+1}(S_{\text{top}}) = W - y_{i+1} - r_{i+1}$.

The euclidean distance from the edge of the next circle to pack C_{i+1} (when positioned at p_{i+1}) and C_j is denoted by $\delta_{i+1}(j)$ and is computed as follows:

$$\delta_{i+1}(j) = \sqrt{(x_{i+1} - x_j)^2 + (y_{i+1} - y_j)^2} - (r_{i+1} + r_j) \qquad (6)$$

The MLDP of the circular piece C_{i+1} when placed at $p_{i+1} \in P_{I_i}$ is calculated as follows:

$$\hat{\delta}_{p_{i+1}} = \min_{\alpha \in I_i \cup E_{\text{edge}} \backslash T_{p_{i+1}}} \{\delta_{i+1}(\alpha)\} \qquad (7)$$

Equation (7) gives the MLDP of C_{i+1} which is computed by using the distances between the piece to place at position p_{i+1} and the elements of the set $I_i \cup \{S_{\text{left}}, S_{\text{bottom}}, S_{\text{top}}\} \backslash T_{p_{i+1}}$ containing the pieces already placed, the three edges of the strip, but by excluding the two elements of $T_{p_{i+1}}$ used for computing the coordinates of C_{i+1} because the corresponding distance is always equal to zero. Note however that the MLDP is equal to zero when C_{i+1} touches more than two elements because one of the three elements does not belong to the set $T_{p_{i+1}}$ and then the distance to this element can be taken into account. Fig. 3 indicates the MLDP $\hat{\delta}_{p_3^{(k)}}$ of each position $p_3^{(k)}$, $k = 1, 2, 3$.

For calculating a packing of the pieces when using the MLDP procedure, the following process is executed: MLDP starts by placing the first circular piece C_1 at the bottom-left corner (at coordinates (r_1, r_1)), the $n-1$ remaining pieces are successively packed by using the MLDP rule as explained above. For example, in Fig. 3, C_3 will be placed at position $p_3^{(1)}$ since the corresponding MLDP has the minimum value.

4   Beam Search-Based Algorithms for SPP

Akeb *et al.* [19] proposed an augmented beam search algorithm, denoted by SEP-MSBS, for the strip packing problem. SEP-MSBS combines two main techniques:

- A strategy based on the use of separate beams that aims to diversify the search space compared to the standard beam search,
- A restarting strategy that consists to rerun the search by changing the first circle to place. The objective of this second technique is to escape from local optima.

The separate-beams mechanism is displayed in Fig. 4. The root node η_1 at level $\ell = 1$ contains the starting configuration (one circle placed in the bottom-left corner of the rectangle) as well as the possible positions for the $n - 1$ remaining circles. The best positions (having the smallest MLDP values) are chosen for branching, this creates the second level $\ell = 2$ (note that each branching consists to choose a position where to place the next circle). From this second level, separate beams are initiated. More precisely, a beam of width $\omega = 1$ is initiated from the first node (the best node), a beam of width $\omega = 2$ is initiated from the second best node, and so on. Thus, the node at position i in the second level is explored by applying a beam search of width $\omega = i$. This is to say that the best nodes do not require an extensive search, the beam width has then a small value, unlike the last nodes in the level that need larger values for the beam width. The separate-beams strategy was shown in [19] to be better than the standard beam search.

Even if SEP-MSBS obtains good results (often the best results in the literature) on the instances used, its run time remains too large. This is mainly due to the restarting strategy, which is executed m times (the number of different circles (radii) in the instance).

5   An Improved Algorithm for SPP

In this paper, we try to improve the SEP-MSBS algorithm by adding a look-ahead strategy. The look-ahead-based mechanism will be described in Section 5.1. The proposed improved algorithm, denoted by IA, will be given and explained in Section 5.2. Some adjustments are introduced in algorithm IA in order to reduce the computation time, these adjustments concern the number of corner positions to explore by the look-ahead strategy as well as the number of circles to take into account in the restarting strategy.

5.1   A Look-Ahead Based Algorithm

Algorithm SEP-MSBS [19] selects, at each level of the tree, the best nodes by using the MLDP rule (Sect. 3.2). This can be assimilated to a *local evaluation*, the packing process does not take into account the remaining circles to place. The look-ahead proceeds differently. Indeed, given the set of nodes $B = \{\eta_\ell^1, ..., \eta_\ell^\omega\}$ of the current level ℓ in the tree, each node η_ℓ^i is characterized by the set I_{ℓ_i} of ℓ circles already placed in the current rectangle and the set P_{ℓ_i} of corner positions for the remaining circles, the look-ahead evaluates each position $p \in P_{\ell_i}$ by continuing the placement of the remaining circles by using the MLDP rule. The objective is to compute final solutions which will help to choose the actual positions for branching from the current level ℓ. This strategy is implemented in the Look-Ahead Branching Procedure (LABP) displayed in Algorithm 2.

In addition to the set of nodes B, LABP (Algorithm 2) receives as input parameter an indicator `feasible` set to the value `false` as well as a real number $0 < \psi \leq 1$. Parameter ψ serves to determine the proportion of corner positions to evaluate by the look-ahead, for example, if $\psi = 0.8$, then only the best 80% of corner positions (those having the smallest MLDP values) are evaluated. The objective of this parameter is to accelerate the algorithm for large instances (those containing a large number of circles, and therefore a large number of corner positions at each step).

The set Π of positions to evaluate by the look-ahead, as explained above, is constructed in Steps 2 and 3. After that, LABP considers each position $p_j \in \Pi$ (Step 4) by packing the corresponding circle in p_j (Step 5). This generates a new node $\eta_{\ell+1}$ that is added to the set of offspring nodes B_ω. The new node is then processed by placing the remaining circles by using the MLDP rule (Step 6). Two cases may then be distinguished:

– A feasible packing is obtained (Step 7), meaning that the n circles were successfully placed inside the current rectangle. In this case, the procedure stops with `feasible=true` (Steps 8 and 9), meaning that the length L of the rectangle could be decreased;
– A feasible packing was not obtained (the n circles cannot be placed into the current rectangle by MLDP). In this second case, the procedure assigns to the node $\eta_{\ell+1}$ the density of the circles placed (Step 11). The density of a given packing is equal to the sum of the surfaces of the circles placed divided by the surface of the rectangle $L \times W$.

Finally, when all the corner positions are processed without obtaining a feasible packing, then the ω best nodes (those that have led to the highest densities) are returned (Steps 14, 15). This means that the current length of the rectangle is too small and should be increased.

Note that procedure LABP (Algorithm 2) is called by a beam search algorithm denoted by BSLA (Algorithm 3, Line 10). BSLA implements a width-first beam search. It uses an interval search $[\underline{L}, \overline{L}]$ in order to compute the best length of the rectangle containing all the circles. BSLA receives several input parameters: the starting node η_ℓ containing the starting configuration, the beam width value ω, the values of the interval search, and parameter ψ indicating the proportion of corner positions to process by LABP.

BSLA calls, at Step 10, the LABP procedure (Algorithm 2) for each value of the rectangle's length L^*. If LABP has computed a feasible packing with the current value of L, then the best length (L_{best}) is updated (Step 12) and

---

Require: A set $B = \{\eta_\ell^1, ..., \eta_\ell^\omega\}$ of ω nodes, a boolean indicator `feasible=false`, and $0 < \psi \leq 1$

Ensure: A feasible solution if `feasible=true`, or a set B_ω of ω nodes (those leading to the highest densities through the MLDP packing procedure).

1: Let P_{ℓ_i} denotes the set of corner positions of node $\eta_\ell^i \in B$;
2: Let Π be the set of all corner positions of B, i.e., $\Pi = \bigcup P_{\ell_i}$;
3: Reduce Π to the $\lceil \psi \times |\Pi| \rceil$ best corner positions (having the best MLDP values);
4: **for all** corner positions $p_j \in \Pi$ **do**
5:    Pack $C_{\ell+1}$ in p_j and insert the resulting node $\eta_{\ell+1}$ into B_ω;
6:    Place in $\eta_{\ell+1}$ the remaining circles by using the MLDP packing procedure;
7:    **if** all circles are placed **then**
8:       `feasible = true`;
9:       **exit** with a feasible solution;
10:   **else**
11:      Assign to $\eta_{\ell+1}$ the density obtained by MLDP;
12:   **end if**
13: **end for**
14: Reduce B_ω to the ω nodes that led to the highest densities by MLDP;
15: **return** B_ω.

Algorithm 2. The Look-Ahead Branching Procedure (LABP)

the upper bound of the interval search $\overline{L}$ is set equal to the current length. Otherwise, level ℓ is incremented by 1, the best expanded nodes returned by LABP (Step 10) replace the nodes of the current level in the tree ($B = B_\omega$), and B_ω is reset to the empty set (Step 14). If LABP did not succeed to compute a feasible packing with the current value of the rectangle's length (Step 17), then the lower bound of the interval search $\underline{L}$ is set equal to the current value of L, i.e., $\underline{L} = L^*$ (Step 18) meaning that the rectangle's length is too small. Finally, it is to note that the binary interval search is stopped when the difference between $\underline{L}$ and $\overline{L}$ becomes less than or equal to a given gap δ.

5.2   The Improved Algorithm (IA)

The improved algorithm, denoted by IA, is given in Algorithm 4. It combines three main techniques: separate beam search, a restarting strategy, and look-ahead. Fig.4 shows how algorithm IA works.

IA receives as input parameters the beam width ω, parameter τ that serves to indicate the proportion of circles taken into account by the restarting strategy,

---

Require: A node η_ℓ, the beam width ω, the bounds of the interval search $(\underline{L}, \overline{L})$, and $0 < \psi \leq 1$

Ensure: The best value for the rectangle's length (L_{best}) and the corresponding feasible packing.

1: Let B denote the set of nodes to be considered;
2: Let B_ω denote the set of descendants of the nodes in B;
3: Let L_{best} be the best length found so far;
4: Let **feasible** be a boolean indicator;

5: **while** $(\overline{L} - \underline{L} > \delta)$ **do**
6:     Set $B = \{\eta_\ell\}$, where η_ℓ is a starting node of level ℓ characterized by I_ℓ, $\overline{I}_\ell$, and P_{I_ℓ};
7:     $L^* = (\overline{L} + \underline{L})/2$;
8:     feasible = false;
9:     **while** $(B \neq \emptyset$ and **feasible**=false$)$ **do**
10:        $B_\omega = \text{LABP}(B, \text{feasible}, \psi)$;
11:        **if feasible=true then**
12:            $L_{\text{best}} = L^*$; $\overline{L} = L^*$;
13:        **else**
14:            $\ell = \ell + 1$; $B = B_\omega$; $B_\omega = \emptyset$;
15:        **end if**
16:     **end while**
17:     **if feasible=false then**
18:        $\underline{L} = L^*$;
19:     **end if**
20: **end while**

Algorithm 3. Beam Search Look-Ahead algorithm (BSLA)

Require: The beam width ω, parameters τ and ψ

Ensure: A feasible packing with the best length L_{best} for the strip

1: $L_{\text{best}} = \overline{L}$; $\underline{L} = (\pi \times \sum_{i=1}^{n} r_i^2)/W$;
2: Rank the pieces of N in decreasing value of their radii;
3: Let $\mathcal{T}$ be the set of circle types (different circles in N);
4: Reduce $\mathcal{T}$ by keeping only $\lceil \tau \times |\mathcal{T}| \rceil$ circles;
5: Set $i_{order} = 1$, where i_{order} is the index of the first circular piece of the set $\mathcal{T}$;

6: **while** $(i_{order} \leq |\mathcal{T}|)$ **do**
7:    Generate the node η_1, characterized by I_1, $\overline{I}_1$, and P_{I_1}, by placing the first
      circle $C_{i_{order}}$ inside the current rectangle and let $B = \eta_1$;
8:    Branch out of B and generate the list of offspring nodes B_ω;
9:    Let $B = \min(\omega, |B_\omega|)$ nodes having the best MLDPs and corresponding to
      distinct corner positions and reset $B_\omega = \emptyset$;
10:   Let η_2 be the node at position ω in B;
11:   feasible = BSLA$(\eta_2, \omega, \underline{L}, \overline{L}, \psi)$;
12:   **if** feasible= true **then**
13:     $\overline{L}$ and L_{best} are updated if a better length is obtained by BSLA;
14:   **end if**
15:   $\underline{L} = (\pi \times \sum_{i=1}^{n} r_i^2)/W$;
16:   $i_{order} = i_{order} + 1$;
17: **end while**

18: **exit** with the best target length L_{best}.

Algorithm 4. The Improved Algorithm (IA)

and parameter ψ used to choose the proportion of corner positions to evaluate by the look-ahead branching procedure LABP (Algorithm 2). The output of algorithm IA is a feasible packing and the corresponding best length of the rectangle L_{best}.

At Step 1 of algorithm IA, the best length L_{best} is set equal to the upper bound of the length $\overline{L}$ which is computed by an Open Strip Generation Solution Procedure (OSGSP$_a$) [28]. The lower bound of the interval search $\underline{L}$ is set equal to the natural lower bound, i.e., $\underline{L} = (\pi \times \sum_{i=1}^{n} r_i^2)/W$ which corresponds to a density equal to 1, this density is of course not possible to obtain because there is always a non-occupied space between the circles and between the circles and the edges of the rectangle. The pieces are then ranked by decreasing value of their radii (Step 2). The set of circle types $\mathcal{T}$ to use in the restarting strategy is constructed in Steps 3 and 4. The index serving to indicate the first circle to place in the bottom-left corner of the current rectangle is initialized in Step 5.

The root node η_1 (cf. Fig. 4) is generated in Step 7. This corresponds to the placement of circle $C_{i_{order}}$ in the bottom-left corner of the rectangle. In Step 8, the list of offspring nodes B_ω is generated. The set B is after that set equal to the ω best nodes of B_ω, this correspond to level $\ell = 2$ in Fig. 4. Since the separate-beams mechanism is used, then only the node at position ω in this level is explored. The node chosen (η_2) is then transmitted to the Beam

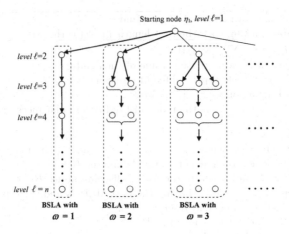

Fig. 4. Separate beams and Look-ahead

Search Look-Ahead algorithm BSLA (Algorithm 3) in order to try to compute a feasible solution (Step 11). If BSLA have reached a feasible packing, then the upper bound $\overline{L}$ of the interval search and the best solution L_{best} are updated (Step 13). Indeed, the upper bound $\overline{L}$ is set to the best value obtained L_{best}. After that the lower bound $\underline{L}$ is reset to the natural lower bound (Step 15). In Step 16, the next circle in set $\mathcal{T}$ is chosen in order to restart the algorithm.

It is to note that the main interest of the look-ahead strategy is that it allows algorithm IA, for which the mechanism is described in Fig.4, to compute feasible solutions from the second level ($\ell = 2$) in the search tree in opposite to the other beam search-based algorithms where feasible solutions are obtained in the last level ($\ell = n$).

6   Computational Results

The algorithms are coded in C++ language and run on a computer with a 3-GHz processor and 256 MB of RAM. Eighteen instances are considered containing from 20 to 200 circles (note that the problem is considered to be large when the number of pieces is at least $n = 100$). The first six instances, denoted by SY1, SY2, SY3, SY4, SY5, and SY6, contain from 20 to 100 circles. They were proposed by Stoyan and Yaskov [14] and are the most known ones in the literature for the strip packing problem, they were for example used in [14], [16], [28], [18], and [19]. Twelve additional instances were proposed by Akeb and Hifi [28], these instances are obtained by concatenating the six original instances of Stoyan and Yaskov and contain from 45 to 200 pieces.

It is to note that all these instances are strongly heterogeneous, i.e., the pieces are practically all of different radii ($m \lesssim n$) where n is the number of circles in the instance and m the number of circle types (different radii).

6.1 Varying the Beam Width When the Look-Ahead Is Used

In a standard beam-search based algorithm, like for algorithm BSBIS [28], it is difficult to know in advance what value to use for the beam width (ω). Indeed, increasing the value of ω does not necessarily improve the solution, even if that increase the search space. This can be explained by the fact that a standard beam search is based on a local evaluation (e.g. MLDP rule) for branching from the current level of the search tree in order to create the next level. As a result, the value of the solution (the length of the rectangle L) oscillates when increasing the beam width. An example is shown in Fig. 5 where BSBIS was executed on instance $SY13$ ($n = 55$, $m = 54$ pieces) for all the values of $1 \leq \omega \leq 30$. Note that this phenomenon concerns also algorithm SEP-MSBS [19] since this one is based on the MLDP selection strategy.

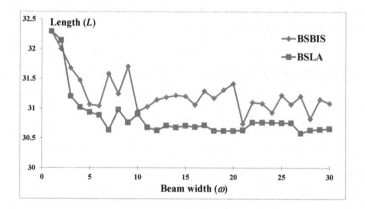

Fig. 5. Comparison between the standard beam search (BSBIS) and BSLA (including look-ahead) on instance SY13 ($n = 55, m = 54$ pieces)

But when the look-ahead is introduced (see Algorithm 3, BSLA), the solution L oscillates much less (as indicated in Fig. 5) and the value of L often decreases when the value of the beam width ω increases. In addition, the solution obtained by the look-ahead (BSLA) is practically always better than that given by BSBIS and the example shown in Fig. 5 is very representative since this phenomenon was shown for all the instances. It is then not necessary to run a look-ahead-based algorithm with all the possible values of ω. In fact, the computational investigation showed that starting with the value $\omega = 10$ and increasing this value by step of 5, i.e., ($\omega = 10 + 5 \times k$, $k \in \mathbb{N}$) corresponds to a good setting. Then, the proposed algorithm (IA) is run with these values of ω.

6.2 Values of Parameters ψ and τ

Another investigation was conducted. It concerns the values of parameter ψ corresponding to the proportion of positions to evaluate by the look-ahead branching procedure (see Algorithm 2) at each level of the tree, and parameter τ that indicates the proportion of circles to use for restarting algorithm IA (see Algorithm 4).

Each parameter ψ and τ was varied in the discrete interval $\{0.5, 0.75, 1\}$, which gives 9 possibilities. The nine possibilities were tested on 3 sets of instances:

- the two smallest instances SY2 ($n = m = 20$) and SY3 ($n = m = 25$),
- two medium-sized instances SY23 ($n = m = 45$) and SY14 ($n = m = 65$),
- two large instances SY6 ($n = 100, m = 98$) and SY1234 ($n = 110, m = 105$).

Table 1 indicates the best values for parameters ψ and τ according to the size of the instance. For example, when considering a small-sized instance ($n < 40$), then all the corner positions have to be processed by the look-ahead ($\psi = 1$) and each circle type have to be used by the restarting strategy ($\tau = 1$). The results of algorithm IA presented in Table 2 and Table 3 are obtained by using the values indicated in Table 1.

Table 1. Best values for parameters ψ and τ according to the size of the instance

Instance size	ψ	τ
small ($n < 40$)	1	1
medium ($40 \leq n < 100$)	0.75	0.75
large ($n \geq 100$)	0.5	1

6.3   Solution Quality of Algorithm IA

Table 2 shows the results obtained by algorithm IA as well as those obtained by different other algorithms. Column 1 (Inst.) contains the name of the instance. Column 2 (n) gives the size of the instance and Column 3 (m) is the number of circle types in the instance. Column 4 (MHD) represents the best length of the rectangle obtained by the Maximum Hole Degree (MHD) heuristic (Huang et al. [16]). The next column (B16) contains the result obtained by a parallel version of MHD (Kubach et al. [18]), symbol "–" means that the result of B16 is not known for the corresponding instances. Column 6 indicates the result obtained by the Beam Search Binary Interval Search algorithm (Akeb and Hifi [28]), the value between parentheses correspond to the value of the beam width with which the solution was obtained. The solution obtained by algorithm SEP-MSBS (Akeb et al. [19]) is given in Column 7 as well as the corresponding beam width. Column 8 (Best Lit.) shows the best known solution in the literature for the studied instances. Finally, the last column contains the result obtained by the Improved Algorithm (IA), the corresponding beam width (ω) is also indicated between brackets. Values in bold characters indicate which algorithm obtains the best solution.

Table 2. Solution quality of algorithm IA

Inst.	n	m	MHD	B16	BSBIS	SEP-MSBS	Best Lit.	IA
SY1	30	30	17.291	17.247	17.2315 (45)	17.2070 (50)	17.2070	**17.0954** (20)
SY2	20	20	14.535	14.536	14.6277 (86)	14.5287 (24)	14.5287	**14.4548** (15)
SY3	25	25	14.470	14.467	14.5310 (78)	14.4616 (44)	14.4616	**14.4017** (80)
SY4	35	35	23.555	23.717	23.6719 (42)	23.4921 (66)	23.4921	**23.3538** (10)
SY5	100	99	36.327	**35.859**	36.0796 (95)	36.1818 (22)	35.8590	36.0045 (15)
SY6	100	98	36.857	**36.452**	36.8456 (85)	36.7197 (26)	36.4520	36.5573 (10)
SY12	50	48	30.067	–	29.7011 (52)	**29.6837** (61)	29.6837	29.7024 (30)
SY13	55	54	30.891	–	30.6371(100)	**30.3705** (68)	30.3705	30.4231 (20)
SY14	65	65	38.265	–	38.0922 (79)	37.8518 (63)	37.8518	**37.6187** (10)
SY23	45	45	28.270	–	27.8708 (98)	**27.6351** (89)	27.6351	27.7148 (35)
SY24	55	54	34.605	–	34.5476 (26)	34.1455 (49)	34.1455	**34.0970** (30)
SY34	60	59	34.901	–	34.9354 (39)	34.6859 (43)	34.6859	**34.5983** (25)
SY56	200	193	69.979	–	64.7246 (65)	65.2024 (06)	64.7246	**64.6904** (10)
SY123	75	72	43.626	–	43.2558 (64)	**43.0306** (25)	43.0306	43.1709 (15)
SY124	85	82	49.335	–	48.8927 (90)	48.8411 (35)	48.8411	**48.6432** (10)
SY134	90	88	49.721	–	49.3954(100)	49.3362 (27)	49.3362	**49.2238** (10)
SY234	80	78	45.888	–	45.9526 (83)	45.6115 (39)	45.6115	**45.4260** (10)
SY1234	110	105	61.906	–	60.2613 (48)	60.0564 (25)	60.0564	**60.0036** (10)

It is to note that the beam-search based algorithms (BSBIS, SEP-MSBS, and IA) were run by using a beam width limit $\bar{\omega} = 100$ and a computation time limit of thirty hours (as in [19]). For a fair comparison, MHD was also run (on the same computer) by using a time limit of thirty hours.

From the results of Table 2, we can see clearly that the new algorithm (IA) has improved twelve results out of eighteen, i.e, 67% of the best known results in the literature. Algorithm SEP-MSBS remains better on four instances (SY12, SY13, SY23, and SY123) and algorithm B16 is better on instances SY5 and SY6.

The computation time is not indicated in Table 2 for algorithm IA because the limit of thirty hours was reached for all the instances except for the smallest one (SY2, $n = m = 20$) for which the algorithm has attained the beam width limit ($\bar{\omega} = 100$) and terminated after 13 hours. For the SEP-MSBS algorithm [19], the time limit was reached for thirteen instances out of eighteen (except for instances SY1, SY2, SY3, SY4 and SY23), i.e., when $n \leq 45$. The reason for which algorithm IA reached the time limit is that the look-ahead strategy consumes a lot of time.

What will be the behavior of the proposed algorithm (IA) when fixing a relatively short time limit? Another investigation, in which the time limit was fixed at thirty minutes, was conducted. Table 3 displays the comparison between the beam search-based algorithms (BSBIS, SEP-MSBS, and IA) when using this new time limit. The first column (Inst.) contains the name of the instance. Column 2 contains the best value obtained by the BSBIS algorithm (based on a standard beam search) as well as the corresponding beam width. Column 3 (t^*) indicates the cumulative computation time (in seconds) in order to obtain the best value

Table 3. Solution quality of algorithm IA when fixing the time limit at 30 minutes

	BSBIS		SEP-MSBS		IA		%imp.	%imp.
Inst.	L	t^*	L	t^*	L	t^*	BSBIS	SEP-MSBS
SY1	17.2315	166	17.2145	1463	**17.2029**	1790	0.17%	0.07%
SY2	14.6277	222	14.5287	155	**14.4548**	216	1.18%	0.51%
SY3	14.5310	308	14.4616	1253	**14.4106**	750	0.83%	0.35%
SY4	23.6719	211	23.5335	1662	**23.3538**	1007	1.34%	0.76%
SY5	36.4042	445	36.3362	1324	**36.1707**	1432	0.64%	0.46%
SY6	36.9387	1637	37.2555	669	**36.9232**	1135	0.04%	0.89%
SY12	**29.7011**	875	30.0447	650	29.9744	1800	-0.92%	0.23%
SY13	30.7415	165	30.7843	1800	**30.6149**	1710	0.41%	0.55%
SY14	38.3573	885	38.2962	851	**37.9690**	1501	1.01%	0.85%
SY23	27.9146	1116	28.0388	885	**27.8493**	1768	0.23%	0.68%
SY24	34.5476	266	34.6732	766	**34.3544**	675	0.56%	0.92%
SY34	34.9354	720	34.9614	1304	**34.7531**	914	0.52%	0.60%
SY56	65.5565	1022	65.7608	1800	**65.3079**	1800	0.38%	0.69%
SY123	43.4907	1745	43.5815	1412	**43.4793**	1511	0.03%	0.23%
SY124	49.3281	456	49.6348	1720	**49.1915**	1661	0.28%	0.89%
SY134	49.8705	1536	49.9136	1397	**49.8184**	1621	0.10%	0.19%
SY234	45.9913	775	46.1901	880	**45.9209**	1321	0.15%	0.58%
SY1234	60.9055	565	60.8783	1800	**60.5660**	1369	0.56%	0.51%

L in Column 2. The results obtained by the two other algorithms (SEP-MSBS and IA) are indicated in Columns 4–7. Column 8 gives the percentage of improvement obtained by the new algorithm IA on BSBIS, the improvement is computed as $\frac{L_{BSBIS} - L_{IA}}{L_{BSBIS}} \times 100\%$. In the same way, the last column contains the percentage of improvement obtained by algorithm IA on algorithm SEP-MSBS.

From Table 3, we can see clearly that when using a relatively short time limit (which is more practical), the proposed algorithm (IA) is practically always the best one (in 17 cases out of 18), except for the instance SY12 where BSBIS remains better. The good results obtained by algorithm IA can be explained by the fact that the look-ahead strategy computes quickly feasible solutions, i.e., from level $\ell = 2$ in the search tree (see Fig. 4) when BSBIS and SEP-MSBS obtain feasible solutions at level $\ell = n$ only. So, even if algorithm IA is stopped after a short computation time, it will have calculated a lot of feasible solutions, increasing the probability to obtain good ones.

Fig. 6 shows the evolution of the best solution obtained by algorithms BSBIS, SEP-MSBS, and IA on instance SY124 (85 circles) when the computation time is limited to thirty minutes (1800 seconds). Algorithm SEP-MSBS is taken as a reference and the x−Axis indicates the cumulative computation time for this algorithm for each value of ω (the beam width). For example, SEP-MSBS needs 108 seconds for a complete run with $\omega = 1$ and 1699 seconds for the five first values of ω. Then after each run of SEP-MSBS with a given value of ω, the

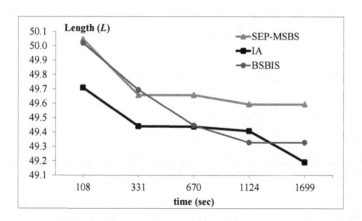

Fig. 6. Evolution of the solution on instance SY124 in the interval of 30 minutes

best length achieved is compared to that obtained by the two other algorithms (BSBIS and IA) for the same cumulative computation time. The results indicate that algorithm BSBIS is better than SEP-MSBS when $\omega > 2$ but within thirty minutes (SEP-MSBS is better on this instance when using a large computation time as indicated in Table 2). Algorithm IA is better than the two others (BSBIS and SEP-MSBS) until $t = 670$ seconds. After that BSBIS achieved a better solution than IA. But IA outperforms BSBIS when $t > 1300$ seconds.

Fig. 7 displays the solution obtained by the proposed algorithm (IA) on the smallest instance (SY2) that contains 20 circles. The new best length is $L = 14.4548$, the previous best known value in the literature was $L = 14.5287$. Fig. 8 shows the new solution obtained by algorithm IA on a medium-sized instance (SY14) that contains 65 circles with $L = 37.6187$. Finally, Fig. 9 displays the solution obtained by algorithm IA on the largest instance (SY56) that contains 200 circles. The new best length is $L = 64.6904$.

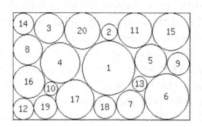

Fig. 7. Solution obtained by the proposed algorithm IA on the smallest instance SY2 ($n = m = 20, L = 14.4548$)

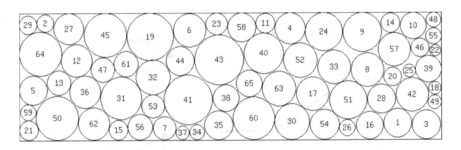

Fig. 8. Solution obtained by algorithm IA on a medium-sized instance SY14 ($n = m = 65, L = 37.6187$)

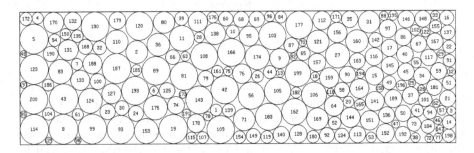

Fig. 9. Solution obtained by algorithm IA on the largest instance SY56 ($n = 200, m = 193, L = 64.6904$)

7 Conclusion

In this paper an improved algorithm, denoted by IA, was proposed in order to solve the strip packing problem. IA is a beam-search based algorithm that includes a look-ahead strategy in order to improve the selection mechanism at each level of the tree. In addition, a restarting strategy was also used.

The computational investigation, conducted on a set of well-known instances in the literature, showed the effectiveness of the proposed algorithm since it has succeeded to improve 67% of the best known solutions in the literature. In addition, another experimentation indicated that the look-ahead obtains good solutions more quickly, i.e., faster than the existing beam-search based algorithms. More precisely, algorithm SEP-MSBS, that does not implement the look-ahead strategy, works well when the computation time is large but its performance decreases when using a relatively short computation time (thirty minutes for example) where algorithm BSBIS is better than SEP-MSBS on more than half of the instances used. The proposed algorithm (IA), thanks to the look-ahead and the optimization of the parameters of this strategy as well as those of the restarting one, achieve good results even for short computation time.

As a future work, it would be interesting to use a parallel algorithm in order to reduce the computation time.

References

1. Baltacioglu, E., Moore, J.T., Hill, R.R.: The distributor's three-dimensional pallet-packing problem: a human intelligence-based heuristic approach. Int. J. Oper. Res. 1, 249–266 (2006)
2. Bortfeldt, A., Homberger, J.: Packing first, routing seconda heuristic for the vehicle routing and loading problem. Comput. Oper. Res. 40, 873–885 (2013)
3. Castillo, I., Kampas, F.J., Pintér, J.D.: Solving circle packing problems by global optimization: Numerical results and industrial applications. Eur. J. Oper. Res. 191, 786–802 (2008)
4. Conway, J.H., Sloane, N.J.A.: Sphere packings, lattices and groups. A Series of comprehensive studies in Mathematics, vol. 290, 703 pages. Springer (1999)
5. Lewis, R., Song, S., Dowsland, K., Thompson, J.: An investigation into two bin packing problems with ordering and orientation implications. Eur. J. Oper. Res. 213, 52–65 (2011)
6. Menon, S., Schrage, L.: Order allocation for stock cutting in the paper industry. Oper. Res. 50, 324–332 (2002)
7. Wäscher, G., Haussner, H., Schumann, H.: An improved typology of cutting and packing problems. Eur. J. Oper. Res. 183, 1109–1130 (2007)
8. Huang, W.Q., Li, Y., Li, C.M., Xu, R.C.: New heuristics for packing unequal circles into a circular container. Comput. Oper. Res. 33, 2125–2142 (2006)
9. Hifi, M., M'Hallah, R.: A dynamic adaptive local search algorithm for the circular packing problem. European J. Oper. Res. 183, 1280–1294 (2007)
10. Akeb, H., Hifi, M., M'Hallah, R.: A beam search based algorithm for the circular packing problem. Comput. Oper. Res. 36, 1513–1528 (2009)
11. Graham, R.L., Lubachevsky, B.D., Nurmela, K.J., Östergård, P.R.J.: Dense packings of congruent circles in a circle. Discrete Math. 181, 139–154 (1998)
12. Liu, J., Xue, S., Liu, Z., Xu, D.: An improved energy landscape paving algorithm for the problem of packing circles into a larger containing circle. Comput. Ind. Eng. 57, 1144–1149 (2009)
13. George, J.A., George, J.M., Lamar, B.W.: Packing different-sized circles into a rectangular container. Eur. J. Oper. Res. 84, 693–712 (1995)
14. Stoyan, Y.G., Yaskov, G.N.: Mathematical model and solution method of optimization problem of placement of rectangles and circles taking into account special constraints. Int. Trans. Oper. Res. 5, 45–57 (1998)
15. Hifi, M., M'Hallah, R.: Approximate algorithms for constrained circular cutting problems. Comput. Oper. Res. 31, 675–694 (2004)
16. Huang, W.Q., Li, Y., Akeb, H., Li, C.M.: Greedy algorithms for packing unequal circles into a rectangular container. J. Oper. Res. Soc. 56, 539–548 (2005)
17. Birgin, E.G., Martinez, J.M., Ronconi, D.P.: Optimizing the packing of cylinders into a rectangular container: A nonlinear approach. Eur. J. Oper. Res. 160, 19–33 (2005)
18. Kubach, T., Bortfeldt, A., Gehring, H.: Parallel greedy algorithms for packing unequal circles into a strip or a rectangle. Cent. Eur. J. Oper. Res. 17, 461–477 (2009)
19. Akeb, H., Hifi, M., Negre, S.: An augmented beam search-based algorithm for the circular open dimension problem. Comput. Ind. Eng. 61, 373–381 (2011)
20. Huang, W.Q., Ye, T.: Greedy vacancy search algorithm for packing equal circles in a square. Oper. Res. Lett. 38, 378–382 (2010)

21. Specht, E.: High density packings of equal circles in rectangles with variable aspect ratio. Comput. Oper. Res. 40, 58–69 (2013)
22. Locatelli, M., Raber, U.: Packing equal circles in a square: a deterministic global optimization approach. Discrete Appl. Math. 122, 139–166 (2002)
23. López, C.O., Beasley, J.E.: A heuristic for the circle packing problem with a variety of containers. Eur. J. Oper. Res. 214, 512–525 (2011)
24. Birgin, E.G., Sobral, F.N.C.: Minimizing the object dimensions in circle and sphere packing problems. Comput. Oper. Res. 35, 2357–2375 (2008)
25. Birgin, E.G., Gentil, J.M.: New and improved results for packing identical unitary radius circles within triangles, rectangles and strips. Comput. Oper. Res. 37, 1318–1327 (2010)
26. Ow, P.S., Morton, T.E.: Filtered beam search in scheduling. Int. J. Prod. Res. 26, 35–62 (1988)
27. Akeb, H., Hifi, M.: Adaptive algorithms for circular cutting/packing problems. Int. J. Oper. Res. 6, 435–458 (2009)
28. Akeb, H., Hifi, M.: Algorithms for the circular two-dimensional open dimension problem. Int. Trans. Oper. Res. 15, 685–704 (2008)

Experimental Evaluation of Pheromone Structures for Ant Colony Optimization: Application to the Robot Skin Wiring Problem

Davide Anghinolfi, Giorgio Cannata, Fulvio Mastrogiovanni, Cristiano Nattero, and Massimo Paolucci

Department of Informatics Bioengineering, Robotics and Systems Engineering, University of Genoa, Via Opera Pia 13, 16145, Genoa, Italy
{davide.anghinolfi,giorgio.cannata,fulvio.mastrogiovanni, cristiano.nattero,massimo.paolucci}@unige.it

Abstract. The problem of optimally routing the wiring in large-scale modular skins for robots is gaining much attention in the literature. Theoretically, the problem is NP-hard. On the basis of previous work [3], [37], we solve the skin wiring problem using an Ant Colony Optimization approach. In this Chapter, we address the problem of designing a good pheromone structure: we propose five alternatives, which are validated using both real and artificially generated problem instances.

Keywords: ant colony optimization, robotics, pheromone structures, skin wiring.

1 Introduction

In order to provide humanoid robots with tactile sensing capabilities, the development of *robot skins* has been an active field of research in the past few years [15]. A robot skin is a sensing device composed of a huge number of networked tactile sensors. Robot skins are expected to enable new means of physical human-robot interaction [5].

Different transduction principles are usually exploited, namely pressure, proximity or temperature [15]. However, to design a robot skin is a hard engineering task, since it requires to deal with such conflicting requirements as resolution [42], reaction dynamics and bandwidth [6], weight, energy consumption, optimal placement and calibration [11], as well as reliability and real-time SW performance [10], [47].

The reference robot skin [12], [41] exploits capacitance-based transducers. In the current HW design, up to 12 tactile elements (i.e., *taxels*) are hosted by a triangular module, which is made by flexible Printed Circuit Board (PCB) and hosts also the read-out electronics, as shown in Figure 1a. Each triangular module can be interconnected to up to 3 other triangular modules to cover large robot body parts, thereby forming a *skin patch* (Figure 1b). A patch can be composed of up to $C = 16$ interconnected triangular modules. Each patch

S. Fidanova (Ed.): *Recent Advances in Computational Optimization*, SCI 470, pp. 93–114.
DOI: 10.1007/978-3-319-00410-5_6   © Springer International Publishing Switzerland 2013

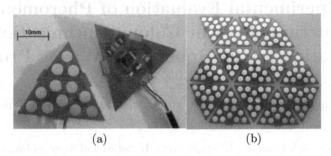

Fig. 1. (a) a skin triangular module PCB, front and rear sides; (b) a parch of inter-connected modules

is managed by an external micro-controller, which is responsible for the trans-mission of tactile sensory data to an embedded PC for further data processing, thereby implementing high-level tactile-based behaviours [9], [16]. The data net-work between triangular modules and the patch micro-controller is embedded within the PCB. Thanks to I/O ports on triangular module sides, tactile sensory data can be sent to and forwarded from adjacent modules, eventually reaching a specific I/O port of the patch connected to the external micro-controller.

In order to reduce the wiring complexity inside the PCB, in the reference technology each micro-controller can be directly connected to one element of a patch only, which is referred to as the *entry point*, whereas the signal cables of other triangular modules are routed via neighbouring modules. Each module can route the signals related to the managed taxels to a single micro-controller. The skin wiring problem consists in finding an appropriate routing of the connections between adjacent triangular modules, in order to optimally link each triangu-lar module to an appropriate micro-controller, possibly satisfying (at the same time) a number of functional requirements, namely *power consumption* and *fault tolerance*. The former is achieved by reducing the number of micro-controllers to the strictly necessary minimum needed to control all of the triangular mod-ules, whereas the latter prescribes that, in case of failure, at least a graceful performance degradation of the robot skin subsystem must be achieved.

The failure of a micro-controller is considered as the most critical situation, since it affects all the associated triangular modules. The failure of a micro-controller can cause large parts of the sensing surface to stop working. Intuitively, as it has been discussed in [3], in order to reduce the effects of such a failure it is advisable to (i) uniformly distribute the load among micro-controllers, and (ii) spread the triangular modules assigned to a micro-controller as much as possible with respect to the robot surface that must be covered.

The concepts above lead to the formulation of skin wiring as an optimization problem with three distinct but interconnected aspects [3]: (i) assign each tri-angular module to a micro-controller; (ii) identify an entry point in the patch for each micro-controller; (iii) define wire routing in terms of interconnections between I/O ports.

A general optimization technique based on Ant Colony Optimization (ACO) has been discussed in [3], [37]. The main contribution of this Chapter is a computational analysis of different pheromone structures for the same ACO algorithm. The Chapter is organized as follows. A graph theoretical problem definition is given in Section 2, followed by a review of relevant literature in Section 3. Section 4 reports a Mixed Integer Programming (MIP) formulation for the problem. We describe a multi-start heuristic and an ACO based approach, together with five different pheromone structures, in Section 5. In Section 6 we discuss the performance of the five heuristic approaches, and we draw conclusions in Section 7.

2 Problem Statement

A robot skin patch can be represented using a graph $G = (V, E)$ where $V = \{v_0, \ldots, v_n\}$ is a set of nodes and $E = \{e_0, \ldots, e_m\}$ is a set of edges linking adjacent triangular modules.

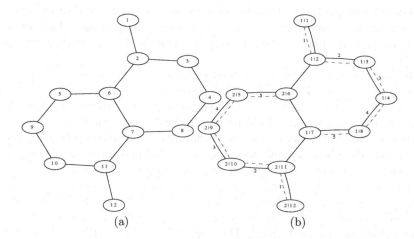

Fig. 2. (a) an example of a graph representing a patch; (b) a possible solution of the associated skin wiring problem

An example of a graph associated with a patch is given in Figure 2a. The graph is symmetric (i.e., no direction is defined between two connected elements) and planar. If border nodes are neglected, then G is also a regular graph. Given a patch and a corresponding graph G, D_{ij} represents the Euclidean distance between the centroids of each pair of triangular modules $(i, j) \in V \times V$. Finally, a set $K = \{1, \ldots, q\}$ of q micro-controllers is given.

The skin wiring problem can be modelled as a *Constrained Spanning Forest* problem: the set of modules assigned to a micro-controller k corresponds to a set of nodes T_k, such that $|T_k| \leq C$ (i.e., the maximum number of modules managed by a micro-controller), and the wiring defines an acyclic connected sub-graph

$S_k(T_k, E_k)$ in G induced by T_k, i.e., a tree. Any node v in T_k can be chosen as root and used as entry point for the micro-controller k. The solution is then a forest F corresponding to the collection of all the trees S_k. The connections of modules in a patch to a set of micro-controllers and the consequent wire routing correspond to a constrained spanning forest on the graph associated with the patch. As an example, Figure 2b shows a possible wiring for the patch graph of Figure 2a using two micro-controllers. In the Figure, node labels $(k|v)$ denote the micro-controller index followed by the node index, whereas the number associated with the edges represents the order of assignment of a connection to a micro-controller.

3   Related Work

The Problem of Wiring in Robot Skin Design. Thanks to the relative design simplicity, research activities on large-scale tactile sensing exploited square regular grid wiring patterns [2], [32], [28], [22], [13], [19], [43], [31]. However, solutions based on grid-like regular patterns are not efficient when scaling-up to large surfaces of robot skin including hundreds or thousands tactile sensors, since the number of wires quadratically increases with the size of the tactile array. To this aim, it is often suggested to introduce multiplexing components, which are able to provide taxel readings at the price of a lower tactile data acquisition rate [30], [38], [21].

The problem of wiring in robot skin design has been addressed in a principled way only in [26], [27], [41], [34], [3]. All of these approaches represent examples of *modular* skin design. In order to build large-scale tactile surfaces, it is necessary to interconnect a (typically) huge number of basic units. Such an interconnection is also used to transmit local tactile data to a centralized processing node. To this aim, each basic module must provide a well-defined signal routing such that, when interconnected with other modules, the overall network is consistent and efficient. The work described in [26], [27] provides a solution for a specific shape of basic modules. Although the solution proves to be efficient, the design is not characterized by any *optimality* principle (i.e., robustness, efficiency, fault-tolerance) leading to good wiring. The same can be said for the work presented in [41], where the wiring between modules is fixed. Self-configuration capabilities as far as wiring is concerned have been implemented in the Hex-O-Skin technology [34]. However, it can not be proved that the obtained configuration is able to enforce any optimality principle. Starting from the robot skin design introduced in [41], we presented heuristic algorithms in [3] to provide a good trade-off between the need for a modular design and (to a limited extent) the requirements associated with robustness and fault-tolerance. In particular, the main proposed algorithm is based on an Ant Colony Optimization (ACO) approach, which we recently improved in [37].

Constrained Spanning Forests. The *Maximum Weight Forest* (MaxWF) problem consists in defining a spanning forest that maximizes the sum of the weights associated with each edge. Although it is possible to solve MaxWF in polynomial-time with a greedy algorithm [39], the introduction of constraints makes the problem much more difficult. The *Maximum Weight t-restricted Forest* (MaxWCF(t))

problem, consisting in finding the MAxWF constrained to have no more than t edges in each tree, has been shown to be NP-hard [23]. Furthermore, the authors give a $1/4$ approximating greedy algorithm. The work in [29] deals with the *Minimum Weighted Constrained Spanning Forest* (MINWCF(p))) problem, which consists in computing the spanning forest of minimum weight such that every tree component contains at least p nodes. In particular, they demonstrate that the problem is NP-hard, $\forall p \geq 4$, whereas in [35] it is shown that MINWCF(p) is NP-hard even with $p = 3$. The work in [7] address the unweighted version of the same problem (MINCF(p)) and show that it is NP-hard for any $p \geq 4$, even on planar bipartite graphs of maximum degree 3. Moreover, they demonstrate that, by dropping the condition of planarity, the problem becomes APX-hard for any $p \geq 3$, even on graphs with maximum degree 3.

Despite a number of similarities, the skin wiring problem differs both from MAxWF and MINWCF (as well as from MINCF). As a matter of fact, the problem considered here is to find a constrained spanning forest F that serves all nodes, minimizing a cost function with two components: (i) the sum, for all the trees $S_k \in F$, of the absolute value of the difference between the number of nodes $|T_k|$ and the desired average number of nodes per tree, defined as $\lambda \triangleq \left\lceil \frac{|V|}{|K|} \right\rceil$ (where $|K|$ is the number of trees), which is expected to favor the micro-controller load balancing; (ii) the sum, for all the trees $S_k \in F$ and for all the ordered pairs of nodes (i, j), such that $i < j$, $i, j \in T_k$, of the difference between the maximum distance between all the pairs of nodes in V and the distance between i and j, which is expected to favor the spreading of triangular modules associated with a single micro-controller.

Considering the second objective component, it is evident that, for the faced *Constrained Spanning Forest* problem, no cost can be *a priori* associated with an edge as it depends on which subsets of nodes in V are finally assigned the $|K|$ trees. As a consequence, since the contribution of an edge to the objective function becomes known only *after* the introduction of that edge in a solution, a greedy heuristic appears not to be suitable for the skin wiring problem. Nonetheless, if the contribution of each edge was known *a priori*, the considered problem could be reduced to a MINWCF. This allows us to conclude that the problem is NP-hard.

Graph Clustering and Partitioning. Graph clustering represents a field of active research [40]. The problem consists in grouping graph vertexes into clusters which consider the graph edge structure in order to maximize the number of edges *within* each cluster and to minimize the number of edges *between* clusters. Although it can be argued that the skin wiring problem belongs to the family of graph clustering problems, typical graph clustering does not explicitly address the connectivity constraints we face in our case (due to the adopted technology) and which makes the problem much harder. The work in [46] is one of the few examples considering such constraints explicitly. However, the authors do not address cardinality constraints, which is a specific requirement in our case.

Graph Partitioning, Packing and Covering [20] consist in grouping the nodes of a graph such that each node is assigned to, respectively, at most or at least one graph subset. These problems are closely related to the topic of this Chapter. However, connectivity is not explicitly enforced in the standard formulation of the problem.

A close class of problems is the Graph Tree Partition (GTP), where the graph *must* be partitioned by means of trees. These problems are NP-hard in many relevant cases, as it has been demonstrated in [14]. In GTP problems, connectivity is ensured by the existence of trees. However, the difference with the problem we discuss in this Chapter resides in the objective and the cardinality constraints. It is worth noting that in [14] the problem is solved by means of an ACO approach.

4 Mathematical Formulation

The skin wiring problem can be formulated as the problem of determining a spanning forest on the graph $G = (V, E)$ associated with a skin patch so that the forest is composed by at most $|K|$ trees and each tree includes at most C nodes. In addition, as for each micro-controller an entry point and a feasible routing must be determined, a root node and an orientation is selected for each tree so that the solution corresponds to a directed spanning forest consisting of a set of arborescences.

The problem can be formulated as a Mixed Integer Programming (MIP) problem considering the directed graph $G' = (V', A')$, where $A' = A \cup \{(s, i) \mid \forall i \in V\}$, $A = \{(i, j), (j, i) : \forall (i, j) \in E\}$ and $V' = V \cup s$. The node s is a dummy node used as a root. In the following, let $a = (i, j)$ denote a generic directed arc, where nodes i and j are respectively the tail and the head of a. Then, a feasible solution identifying a directed spanning forest H is a set of arborescences R_k, i.e., $H \triangleq \{R_k(T_k, A_k), k \in K\}$, where $T_k \subseteq V$ and $A_k \subseteq A$. The following constants are used: $D_{ij}, \forall (i, j) \in V \times V$, is the Euclidean distance between nodes i and j; $D^{\max} = \max\{D_{ij}\}$; $D^{\min} = \min\{D_{ij}\}$; C is the maximum number of nodes in a tree (i.e., the micro-controller capacity, which is 16 in our case); $\lambda \in (0, C]$ is the desired number of nodes in a tree (i.e., the desired micro-controller load level); $w_1, w_2, w_3 \in \mathbb{R}_+$ are the weights modelling the priority of the objective function components.

The following decision variables are used: $r_{kv} \in \{0, 1\}$, $\forall k \in K, \forall v \in V$; $r_{kv} = 1$ if only if controller $k \in K$ is connected directly to node $v \in V$, that is, v is the root of tree k (i.e., the entry point for micro-controller k); $x_{ka} \in \{0, 1\}$, $\forall k \in K, \forall a \in A$; $x_{ka} = 1$ if only if $a \in A_k$; a node j is assigned to an arborescence (T_k, A_k) (i.e., it is controlled by micro-controller k) if either j is the root of k or it is the head of an arc $a = (i, j)$ such that $a \in A_k$; $y_{ij} \in \{0, 1\}$, $\forall (i, j) \in V \times V$, $i < j$; $y_{ij} = 1$ only if $i, j \in T_k$ for a given k (these variables determine if two nodes are assigned to the same arborescence and are used to evaluate the relative distance between nodes); $n_v \in \{0, 1\}$, $\forall v \in V$; $n_v = 1$ only if $v \notin T_k, \forall k \in K$; variables n_v are introduced to relax the node spanning condition, i.e., to model also the case of skin patches that generate graphs structured so that there is no possibility to assign all the nodes to a micro-controller (this point is discussed later in this Section); $\Delta_k \in \mathbb{R}_+, \forall k \in K$; these are deviational variables giving the absolute difference between $|T_k|$ and the average desired load λ for micro-controllers; $u_v \in \mathbb{R}$, $\forall v \in V$; variables used in *sub-tour elimination* constraints.

Skin wiring is a multi-objective problem requiring the minimization of the three following objectives:

$$O_1 \triangleq \sum_{v \in V} n_v, \tag{1}$$

$$O_2 \triangleq \sum_{k \in K} \Delta_k, \tag{2}$$

$$O_3 \triangleq \frac{1}{(D^{\max} - D^{\min})} \sum_{\substack{i,j \in V \\ i<j}} y_{ij} \left(D^{\max} - D_{ij} \right), \tag{3}$$

where O_1 (1) is the sum of unassigned nodes in the current solution, O_2 (2) is the sum over all micro-controllers of the variations from the average desired load λ and O_3 (3) corresponds to the sum of the normalized distances among every pair of nodes assigned to the same arborescence.

The following three scaling factors, namely ν_1, ν_2 and ν_3, are introduced to ensure that $O_h \in [0,1]$, $h = 1,2,3$:

$$\nu_1 \triangleq \frac{1}{|V|}, \quad \nu_2 \triangleq \frac{1}{|K|\lambda}, \quad \nu_3 \triangleq \frac{2}{[|V|(|V|-1)]}. \tag{4}$$

Please note that, since such factors are derived from worst case considerations, the scaled objectives can not reach the extremes of their variation range.

The multi-objective problem can be converted into a scalar minimization problem by combining the three objectives into a single weighted additive objective function, where the values of the weights are provided according to the different preference of the decision maker. As a matter of fact, a lexicographic priority can be imposed between any pair of objectives O_i, O_j, fixing $w_i \gg w_j$. Then, the proposed formulation is as follows:

$$\min \quad w_1 \nu_1 O_1 + w_2 \nu_2 O_2 + w_3 \nu_3 O_3, \tag{5}$$

subject to:

$$\sum_{k \in K} (r_{kj} + \sum_{\substack{a \in A \\ a=(i,j)}} x_{ka}) + n_j = 1, \qquad \forall j \in V, \tag{6}$$

$$x_{ka} \leq \sum_{\substack{f \in A \\ f=(j,i)}} x_{kf} + r_{ki}, \quad \forall i \in V, k \in K, a = (i,l) \in A, \tag{7}$$

$$\sum_{v \in V} r_{kv} \leq 1, \qquad \forall k \in K, \tag{8}$$

$$\sum_{j \in V} \left(r_{kj} + \sum_{\substack{a \in A \\ a=(i,j)}} x_{ka} \right) \leq C, \qquad \forall k \in K, \tag{9}$$

$$u_i - u_j + C \sum_{k \in K} x_{ka} \leq C - 1, \qquad \forall a = (i,j) \in A, \tag{10}$$

$$\sum_{j \in V} \sum_{a \in A} (x_{ka} + r_{kj}) - \lambda \leq \Delta_k, \qquad \forall k \in K, \tag{11}$$

$$\sum_{j \in V} \sum_{a \in A} (x_{ka} + r_{kj}) - \lambda \geq -\Delta_k, \qquad \forall k \in K, \tag{12}$$

$$y_{ij} \geq \sum_{\substack{a_1 \in A \\ a_1 = (h,i)}} x_{ka_1} + r_{ki} + \sum_{\substack{a_2 \in A \\ a_2 = (l,j)}} x_{ka_2} + r_{kj} - 1, \tag{13}$$

$$\forall i, j \in V, k \in K, i < j.$$

Constraints (6) require that each node is at most assigned to a single arborescence (i.e., a micro-controller), being either the entry point for the arborescence or the head of one of its arcs. Constraints (7) ensure that an arc with origin in node i can be assigned to controller k only if another arc assigned to k enters in i or if i is an entry point for k. Constraints (8) impose that for each micro-controller there is at most one entry point. Constraints (9) limit the number of nodes assigned to each micro-controller to C, taking into account the fact that a node j is assigned to a micro-controller k or an arc exists insisting on j and starting from a node assigned to k. Constraints (10) are the sub-tour elimination constraints introduced in [33]. These constraints ensure that each micro-controller is associated with an acyclic connected and directed sub-graph, i.e., an arborescence. Constraints (11) and (12) define the unbalance between the number of nodes assigned to a micro-controller k and the average desired load. Constraints (13) impose $y_{ij} = 1$ if nodes i and j are controlled by the same micro-controller, specifically by the first and the second summation. Variables n_v are necessary to determine feasible solutions for the graphs whose nodes cannot be partitioned into $|K|$ distinct arborescences, i.e., assigned to the available $|K|$ micro-controllers. Such situations may arise in case of patches generating a non connected graph, but also in case of connected graph with particular structures [3].

5   A Solution to the Skin Wiring Problem

5.1   An ACO Algorithm for Optimal Skin Wiring

Since the MIP model *as-is* is almost useless to solve large scale instances [3], a meta-heuristic approach has been developed. In particular, an *Ant Colony Optimization* (ACO) algorithm [17], together with an efficient *Candidate Strategy* (CS), has been implemented in [3]. The ACO algorithm described in this Section shares the same structure of the algorithm introduced by [4], which borrows concepts both from the *Ant Colony System* (ACS) [18] and the *Max-Min Ant System* (MMAS) frameworks [45].

Pseudocode 1. An ACO algorithm for the skin wiring problem

Input: a graph $G = (V, E)$, the number of micro-controllers $|K|$, the desired occupancy
level λ and the number of ants P.

```
1:  F* ← ∅
2:  z_best ← +∞
3:  π ← initPheromone(G)
4:  while termination_condition not met do
5:      for all ant a do
6:          F_a ← buildSolution(G,|K|,λ,π)
7:          if z(F_a) < z_best then
8:              F* ← F_a
9:              z_best ← z(F*)
10:         end if
11:         localPheromoneUpdate(F_a,π)
12:     end for
13:     globalPheromoneUpdate(F*,π)
14: end while
15: return F*
```

The algorithm is outlined in Pseudocode 1. The algorithm requires the graph
G, the number $|K|$ of micro-controllers, the desired occupancy level λ and the
number P of artificial ants. The best solution F^* and the best value z_{best} are initialized, respectively, to an empty set (line 1) and to $+\infty$ (line 2). The pheromone
π is initialized at line 3. The main loop (lines 4–14) consists in making each ant
construct a forest F_a which possibly spans all the nodes in V (line 6), comparing
the solution against the incumbent and eventually save it (lines 8–9). Pheromone
trails are locally updated after each construction (line 11) and globally updated
at the end of the iteration (line 13). The best found solution is finally returned
(line 15).

As discussed in [3], the solution construction proceeds incrementally, since
each ant a constructs a forest F_a one tree T at a time. A tree is constructed by
adding an edge e to T. The edge e is obtained from an element z, which is ex-
tracted out of a set Θ of candidates using information associated with pheromone
trails. The set Θ is generated from scratch before the insertion of each edge (refer
to Section 5.2 and Section 5.3). The set Θ can contain one out of two possible
types of elements, namely edges or nodes. A node v is called *free* if it has not
been assigned to any tree yet. If z corresponds to an edge $e = (i, j)$, either $i \in T$
and j is free, or $j \in T$ and i is free. In the other case, if z corresponds to a
node v, then it is free. For the sake of completeness, any edge e connecting v
to a node in T can be added, since their contribution to the objective function
(5) is equivalent. The construction of the current tree ends when the desired
occupancy level has been reached (i.e., when $|T| = \lambda$) or it is no longer possible
to find an element for which the insertion is feasible (i.e., when $\Theta = \emptyset$).

Note that, for comparison purposes, it is possible to design a randomized
constructive heuristic simply by neglecting the use of pheromone. In this case, the
selection of the next element $z \in \Theta$ to insert in T is randomly performed with a

uniform probability distribution. A *Multi Start Heuristic* (MSH) is then obtained by restarting the construction. Conceptually, this approach corresponds to using one ant only, i.e., to setting $P = 1$ but, in practice, it is convenient to implement a dedicated algorithm. For a fair comparison, the termination condition both of MSH and ACO is the maximum running time.

Differently from MSH, when more alternatives exist, the ant performs the selection in two steps. First, as in the standard ACS, the ant randomly chooses the node selection rule between *exploitation* and *exploration* with probability $q \in [0, 1]$ and $1 - q$, respectively. Let $\pi(z)$ denote the pheromone trail associated with element z. Then, the *exploitation* rule selects z deterministically according to

$$z = \arg \max_{w \in \Theta} \pi(w), \tag{14}$$

and the *exploration* rule selects z according to a selection probability $p(z)$ obtained as

$$p(z) = \frac{\pi(z)}{\sum_{w \in \Theta} \pi(w)}. \tag{15}$$

Following the same approach proposed by [4], the pheromone trails do not depend on the cost function values associated with previously explored solutions. Rather, they vary in an arbitrary range $[\pi_{min}, \pi_{max}]$ such that $\pi_{min} < \pi_{max}$ and $\pi_{min} \geq 0$. In this way, π_{min} and π_{max} are no longer parameters to be tuned. This feature makes the algorithm independent from the specific problem or instance. After an ant a completes the construction of a solution F, in order to reduce the probability that the successive ants construct an identical solution during the same iteration, the pheromone trails are locally updated (line 11 in Pseudocode 1) as follows:

$$\pi(z) \leftarrow (1 - \rho)\,\pi(z), \quad \forall z \in F, \tag{16}$$

where $\rho \in [0, 1]$ is the local evaporation parameter. The *global pheromone update* (line 13) is performed in three steps: (i) perturbations due to the local pheromone updates are removed; (ii) pheromone trails are evaporated as follows:

$$\pi(z) \leftarrow (1 - \alpha)\,\pi(z), \quad \forall z \notin F^*, \tag{17}$$

where F^* is the incumbent solution and $\alpha \in [0, 1]$ is the global evaporation parameter; (iii) pheromone trails relevant to F^* are reinforced as follows:

$$\pi(z) \leftarrow \pi(z) + \alpha \cdot \Delta\pi(z), \quad \forall z \in F^*, \tag{18}$$

where $\Delta\pi(z)$ is the maximum pheromone reinforcement obtained as

$$\Delta\pi(z) = \pi_{max} - \pi(z), \quad \forall z \in F^*. \tag{19}$$

As shown in [4], these rules make the pheromone reach both bounds asymptotically.

5.2   A Candidate Strategy

The set Θ of candidate elements for insertion is constructed, coherently with the used pheromone structure, upon a set Ψ of candidate nodes, which is defined as follows. Let $A(v)$ be the set of nodes adjacent to v and $N(v) \subseteq A(v)$ be the set of free nodes adjacent to v. With a minor abuse in the notation, let also $A(T)$ be the sets of nodes adjacent to those in T and $N(T)$ the set of free nodes adjacent to those in T. Given a incomplete forest F, possibly empty, a first set Ψ_0 is obtained including all nodes in $N(T)$. If the tree is empty, then any free node is added to Ψ_0 to bootstrap its construction. The set Ψ_0 can be passed directly to an ant for construction but this choice leads to poor quality solutions which, most of the times, are also incomplete. A first improvement is obtained by constructing a set $\Psi_1 \subseteq \Psi$ keeping only those nodes for which $|N(v)|$ is minimal, i.e.,

$$\Psi_1 = \left\{ v \in \Psi : v = \arg \min_{w \in \Psi} |N(w)| \right\}. \tag{20}$$

This is called the *Least Unassigned* (LU) rule and allows us to obtain much better solutions, although it still produces too many incomplete ones. Finally, a set $\Psi_2 \subseteq \Psi_1$ can be defined by keeping only those nodes in Ψ_1 for which, in turn, the total number of free adjacent nodes is minimal, i.e.,

$$\Psi_2 \triangleq \left\{ v \in \Psi_1 : v = \arg \min_{w \in \Psi_1} \sum_{u \in N(w)} |N(u)|. \right\} \tag{21}$$

This is called the *Least Cumulative Unassigned* (LCU) rule and, thanks to its look-ahead effect, it reduces the chances that a node v is left unassigned to any micro-controller since it tends to avoid leaving v isolated. It is convenient to set $\Psi = \Psi_2$: LCU allows to obtain definitely better solutions in terms of objective O_1, and partly of O_3 since, in most cases, this mechanism tends to produce chain-like shaped trees, i.e., trees with only two leaves. Objective O_2 is implicitly taken into account by limiting the number of nodes per tree to λ.

5.3   Pheromone Structures

The pheromone allows the artificial ants to learn what the attributes of a good solution are. The candidate set must be coherent with the choice. To better explain the structures we explore, throughout the remainder of this Section we consider the scenario depicted in Figure 3, which represents part of a graph where a tree is under construction. Specifically, Figure 3a shows a case where only node 1 is assigned to the tree, whereas in Figure 3b also nodes 2 and 3 have been added. Since in this problem it is necessary to learn how nodes link together, we have tested the following possibilities.

Direct Edges (DE). In this case, Θ is defined as the set of edges linking a candidate node and a node in current tree, i.e.,

$$\Theta_{\mathrm{DE}} \triangleq \{ e \in E, e = (v, i) \vee e = (i, v) : v \in \Psi \wedge i \in T \}. \tag{22}$$

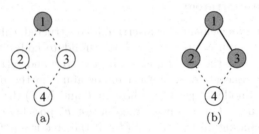

Fig. 3. Two partial solutions for the same graph $G = (V, E)$, where $V = \{1; 2; 3; 4\}$ and $E = \{(1, 2); (1, 3); (2, 4); (3, 4)\}$

Fig. 4. A Direct Edges structure where (a) $\Theta_{\mathrm{DE}} = \{(1, 2); (1, 3)\}$, $\pi = \{\pi_{12}; \pi_{13}\}$ and (b) $\Theta_{\mathrm{DE}} = \{(2, 4); (3, 4)\}$, $\pi = \{\pi_{24}; \pi_{34}\}$

The pheromone trail is associated with each edge $e \in E$ and is indicated as π_e. Figure 4 shows the pheromone and the candidates.

Fig. 5. A Cumulative Edges structure where (a) $\Theta_{\mathrm{CE}} = \{2; 3\}$, $\pi = \{\pi_2, \pi_3\}$, $\pi_2 = \pi_{12}$, $\pi_3 = \pi_{13}$ and (b) $\Theta_{\mathrm{CE}} = \{4\}$, $\pi = \{\pi_4\}$, $\pi_4 = E(\pi_{24}, \pi_{34})$

Cumulative Edges (CE). In this case we have:

$$\Theta_{\mathrm{CE}} = \Psi, \tag{23}$$

and the value of the corresponding pheromone trail π_v is averaged over all possible edges e linking v to a node in T, i.e.,

$$\pi_v \triangleq \frac{\sum_{e \in E_T(v)} \pi_e}{|E_T(v)|}, \tag{24}$$

where $E_T(v) \subseteq E$ is defined as:

$$E_T(v) \triangleq \{e \in E, e = (v,i) \vee e = (i,v) : v \in \Psi, i \in T\}. \tag{25}$$

This is the pheromone structure originally adopted by the authors in [3] and is shown in Figure 5.

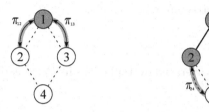

Fig. 6. A Direct Pairs structure where (a) $\Theta_{\mathrm{DP}} = \{(1,2); (1,3)\}$, $\pi = \{\pi_{(1,2)}; \pi_{(1,3)}\}$ and (b) $\Theta_{\mathrm{DP}} = \{(1,4); (2,4); (3,4)\}$, $\pi = \{\pi_{(1,4)}; \pi_{(2,4)}; \pi_{(3,4)}\}$

Direct Pairs (DP). In this case Θ is the set of (virtual) edges linking all nodes in T with all nodes in Ψ, i.e.,

$$\Theta_{\mathrm{DP}} \triangleq \{e \in P_T(v)\}, \tag{26}$$

where

$$P_T(v) \triangleq \{e = (i,j), i < j : (i \in \Psi \wedge j \in T) \vee (i \in T \wedge j \in \Psi)\} \tag{27}$$

The pheromone is associated with the same edges. Figure 6 shows how this structure works.

Fig. 7. A Cumulative Pairs structure where (a) $\Theta_{\mathrm{CP}} = \{2; 3\}$, $\pi = \{\pi_2, \pi_3\}$, $\pi_2 = \pi_{(1,2)}$, $\pi_3 = \pi_{(1,3)}\}$ and (b) $\Theta_{\mathrm{CP}} = \{4\}$, $\pi = \{\pi_{[(1,4);(2,4);(3,4)]}\}$

Cumulative Pairs (CP). This structure behaves as in the CE case, but is applied to pairs defined as in the previous case, i.e.,

$$\Theta_{\mathrm{CP}} = \Psi, \tag{28}$$

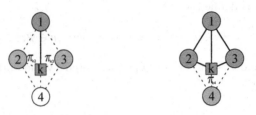

Fig. 8. A Naive Clustering structure where (a) $\Theta_{NC} = \{2; 3\}$, $\pi = \{\pi_{(k,2)}; \pi_{(k,3)}\}$ and (b) $\Theta_{NC} = \{4\}$, $\pi = \{\pi_{(k,4)}\}$

and the pheromone trail is averaged over all edges in $P_T(v)$ ending or beginning in v, i.e.,

$$\pi_v \triangleq \frac{\sum_{e \in P_T(v)} \pi_e}{|P_T(v)|}. \tag{29}$$

The behavior of this structure is further explained in Figure 7.

Naive Clustering (NC). In this case, the element to be inserted is a node, i.e.,

$$\Theta_{NC} \triangleq \Psi, \tag{30}$$

and the pheromone π_c is associated with the ordered couple $c = (T, v)$, where T is the tree under construction and $v \in \Theta_{NC}$. This representation is the only one which works also with empty trees, whereas the others require the adoption of dummy edges to bootstrap the construction. The Naive Clustering is depicted in Figure 8.

6   Experimental Analysis

Generation of Problem Instances. We consider two sets of instances, which we refer to as i and s. Dataset i contains real instances corresponding to patches for body parts of the *iCub* robot platform[1], which have a number of nodes ranging from 6 to 232. These instances correspond respectively to: i1 = left upper arm, lower part; i2 = left hip; i3 = right upper forearm; i4 = upper torso; i5 = lower torso. Dataset s contains artificially generated instances with a number of nodes ranging from 35 to 2470. These instances are identified by prefix s followed by the **number** of nodes. The topology of these instances has the same structure of the real ones, but their size is greater. They have been generated by considering a grid of tactile elements which has been randomly cut using a square shape. We compute the number q of micro-controllers for s instances as $q = \lceil \frac{n}{C} \rceil$, where n is the number of nodes, C is the micro-controller capacity

---

[1] Please refer to the official iCub website at www.icub.org.

and $\lceil$ $\rceil$ denotes the *ceiling* operator. In order to obtain a complete solution for i instances, we repeatedly execute the ACO algorithm, iteratively increasing the number of micro-controllers q.

Experimental Setup and Requirements. In the performed experiments, we compare the pheromone configurations described in Section 5.3 for the ACO algorithm, together with the MSH as comparison term, against datasets i and s. We use a maximum time of 300 seconds as a termination condition. Since algorithms are randomized, we execute 10 independent replications for each instance. All tests are executed on a 2.83 GHz Intel Core 2 Quad CPU $Q9550$, with 4GB of RAM, running Linux (Ubuntu with 3.2.0-26-generic-pae, 32 bit).

We purposively do not report the results obtained using mathematical programming because in previous experiments [3] verified that the mathematical programming model does not scale, and here we are interested in addressing even larger instances.

As a matter of fact, in realistic robot applications, solutions which do not assign all the skin modules to micro-controllers (1) are considered unacceptable, whereas *micro-controller load balancing* (2) is considered more important than *skin module spreading* (3). This can be modelled enforcing a lexicographic ordering $O_1 \prec O_2 \prec O_3$, where $a \prec b$ means that a has a higher priority than b. To model this priority, the weights in the objective function (5) must be set in order to satisfy $w_1 \gg w_2$ and $w_2 \gg w_3$. In particular, the following values are considered suitable for the objective weights: $w_1 = 1000$, $w_2 = 10$, $w_3 = 1$. Although all of the described algorithms explicitly enforce the required lexicographic ordering of solutions, it is noteworthy that this assignment allows for the comparison with the solutions obtained with a MIP solver by [3], which can be used as a ground truth.

Results and Discussion. Table 1 summarizes the performance of the algorithms. The first four columns show, respectively, the instance ID, the number of nodes $|V|$, the number of micro-controllers $|K|$ and the desired occupancy level λ. The fifth column reports the value BV of the best solution found during the experiments, whereas the subsequent columns report the average *Relative Percent Deviation* (RPD) of MSH and the ACO algorithms with the different pheromone configurations (DE, CE, DP, CP and NC). Finally, the last two columns report the range between the worst and the best solution found respectively by all the algorithms and by ACO algorithms only. Each row is dedicated to a single instance, and the best value is highlighted in boldface. The last two rows show the average and the standard deviation, respectively. The RPD is computed as

$$\text{RPD} = \frac{z - \text{BV}}{\text{BV}} \cdot 100, \tag{31}$$

where z is the value of the objective function found by the algorithm under test.

Table 1. Instances characteristics and computational results

ID	$|V|$	$|K|$	λ	BV	MSH	DE	CE	ACO		NC	range (all) (max - min)	range (ACO) (max - min)
								DP	CP			
i1	6	2	3	0.3851	**0.000%**	**0.000%**	**0.000%**	**0.000%**	**0.000%**	**0.000%**	0	0
i2	9	1	9	0.6792	**0.000%**	**0.000%**	**0.000%**	**0.000%**	**0.000%**	**0.000%**	0	0
i3	37	4	13	3.1087	**0.000%**	**0.000%**	**0.000%**	**0.000%**	**0.000%**	**0.000%**	0	0
i4	125	9	16	1.4905	0.023%	0.032%	0.033%	**0.022%**	0.030%	0.035%	0.01	0.01
i5	232	15	16	0.3899	0.240%	0.018%	**0.010%**	0.165%	0.171%	0.215%	0.23	0.21
s35	35	3	12	0.5193	**0.000%**	**0.000%**	**0.000%**	**0.000%**	**0.000%**	**0.000%**	0	0
s40	40	3	14	0.7179	**0.000%**	**0.000%**	**0.000%**	**0.000%**	**0.000%**	**0.000%**	0	0
s54	54	4	14	0.5406	**0.000%**	**0.000%**	**0.000%**	**0.000%**	**0.000%**	**0.000%**	0	0
s60	60	4	15	0.1798	0.006%	0.217%	**0.000%**	1.385%	0.063%	0.063%	1.38	1.38
s77	77	5	16	0.5219	0.093%	**0.000%**	0.076%	0.678%	**0.000%**	0.630%	0.68	0.68
s84	84	6	14	0.1230	2.069%	**0.000%**	**0.000%**	**0.000%**	**0.000%**	2.107%	2.11	2.11
s104	104	7	15	0.2027	0.768%	0.320%	0.360%	0.332%	**0.000%**	1.962%	1.96	1.96
s112	112	7	16	0.1063	1.834%	0.327%	1.906%	0.966%	**0.105%**	0.121%	1.8	1.8
s135	135	9	15	0.0840	4.900%	**0.000%**	**0.000%**	4.035%	1.527%	4.336%	4.9	4.34
s170	170	11	16	0.4137	0.482%	0.150%	**0.012%**	0.521%	0.283%	0.271%	0.51	0.51
s198	198	13	16	0.5440	0.263%	0.150%	0.125%	0.079%	**0.052%**	0.216%	0.21	0.16
s252	252	16	16	0.2070	0.730%	0.199%	**0.150%**	0.349%	0.357%	0.372%	0.58	0.22
s322	322	21	16	0.4570	0.201%	0.091%	**0.079%**	0.126%	0.175%	0.178%	0.12	0.1
s432	432	27	16	0.0312	1.970%	**0.133%**	0.511%	0.705%	0.622%	1.152%	1.84	1.02
s608	608	38	16	0.0225	2.230%	**0.309%**	0.914%	1.205%	0.718%	1.128%	1.92	0.9
s874	874	55	16	0.0840	0.324%	**0.041%**	0.113%	0.128%	0.154%	0.135%	0.28	0.11
s1344	1344	84	16	0.0105	1.125%	0.366%	**0.106%**	0.381%	0.665%	0.110%	1.02	0.56
s2470	2470	155	16	0.0461	0.082%	0.018%	**0.013%**	0.030%	0.044%	0.035%	0.07	0.03
average					0,754%	**0,103%**	0,192%	0,483%	0,216%	0,568%		
stdev					1,174%	**0,127%**	0,431%	0,876%	0,364%	1,024%		

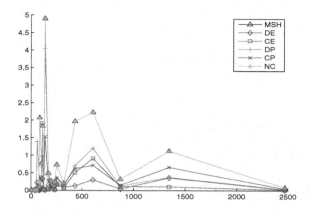

Fig. 9. Algorithms scaling capability with size

The same values reported in Table 1 are visualized in Figure 9, which plots the results as a function of the number of nodes.

First of all, although not specifically highlighted, it is remarkable that every algorithm is able to produce a complete solution, i.e., with $O_1 = 0$, in every run on every instance. This is a very satisfying performance from the point of view of the subject matter experts. From Table 1 it stems out that every algorithm proves to be very efficient on the less challenging instances, namely i1-i3 and s35-s54: not only 0% RPD is achieved for all of them, but they actually keep finding a solution with the same (best) cost in all of the replications. In Figure 9, this corresponds to those points, on the left, where all the lines reach the horizontal axis. At this stage of the analysis, it is not possible yet to state that there are algorithms dominated by others, although the ACO with the Direct Edges pheromone structure achieves the best average results and the tightest standard deviation. Considered on average, the remaining algorithms rank as CE, CP, DP, NC and MSH, whereas looking at the standard deviation the order is DE, CP, CE, DP, NC and DE. The right part of Figure 9 reinforces the observation that DE, CE and CP perform better on large scale instances. Figure 10 show the average RPD of the various algorithms as a function of running time.

To gain more insight, it is useful to filter out the simplest instances and compute the confidence intervals. Table 2 shows, on the rows, the average RPD (avg) of each algorithm, together with the 95% confidence intervals (lo and up, respectively) and the standard deviation (stdev). For convenience, the same intervals are also plotted in Figure 11.

The confidence interval of DE does not overlap with that associated with MSH, which allows us to conclude that the improvement obtained with the Direct Edge pheromone representation is statistically significant. Adding this learning mechanism to the purely random constructive mechanism adopted by the MSH allows for better results. Interestingly, DE is also better than NC, for which the interval is largely coincident with that of MSH, and this is remarkable:

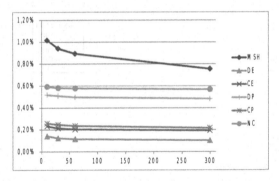

Fig. 10. Average RPD% as a function of running time

Table 2. 95% confidence intervals

	lo	avg	up	stdev
MSH	0.42%	1.02%	1.62%	1.27%
DE	0.08%	0.14%	0.20%	0.13%
CE	0.03%	0.26%	0.49%	0.49%
DP	0.19%	0.65%	1.11%	0.97%
CP	0.10%	0.29%	0.48%	0.40%
NC	0.23%	0.77%	1.31%	1.13%

apparently, the clustering information associated with the couple $c = (T, v)$, where T is a tree and v is a node is not powerful enough. The reason cannot depend on the choice of the edge to be inserted in T because, given node v, the contribution to the objective function is uniquely defined. To understand one possible reason why NC does not perform well, let us consider a graph for which the set of nodes can be partitioned into two disjoint subsets V_1 and V_2, and suppose that it is possible to obtain a solution for which V_1 and V_2 are assigned to trees T_1 and T_2 respectively. Then, a completely equivalent solution can be obtained by swapping the assignments, i.e., by assigning V_2 to T_1 and V_1 to T_2. This kind of representation suffers from the inefficiency due to its embedded strong symmetry: since equivalent solutions can be obtained simply by performing cyclic permutations of the assignments represented in couples c, the algorithm is likely to waste a lot of CPU time to determine which of these equivalent assignments works best. It can be argued that, at least on the given instances, it is not efficient to represent the clustering information by specifying the tree to which a node must be assigned to. Instead, the remaining structures try to represent the fact that two or more nodes must be grouped together in a good solution. The performance of DE is superior also to that of DP: even if there is a minimal intersection between the two, the interval of DP is almost entirely contained into that of NC. DE, CE and CP are the best configurations. The interval of DE is embedded into that of CE and overlaps for almost its 80% with that of CP, and therefore no statistical significance of the superiority of DE can be inferred.

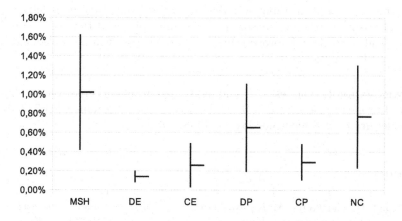

Fig. 11. Average results for MSH and ACOs with 95% confidence intervals

As discussed above, DE, CE and CP are the three best configurations. At a first glance, it is possible to argue that the difference in performance of the three methods vanishes when the complexity scale of the instances grows. As long as problem instances become large (i.e., considering a number of nodes approximately exceeding 1400), the three structures provide mostly equivalent results. On the one hand, if a single pheromone structure had to be selected, then DE would be a good choice, as it is the only one guaranteeing superior performance with respect to that offered by MSH. On the other hand, CE and CP would be the next best alternatives. Summarizing, these comments allow us to speculate about potentially *good features* of pheromone structures, namely (i) the use of topology, and (ii) the information about the fact that two or more elements (i.e., nodes in our case) must belong to the same cluster. It emerges that DE and CE are characterized by both the features, whereas CP is characterized by just the second one.

7   Conclusions

In this Chapter, five different pheromone structures for an Ant Colony Optimization algorithm have been designed and tested to solve a minimum cost *Constrained Spanning Forest* problem arising in robot skin design.

The proposed heuristics, namely Direct Edges (DE), Cumulative Edges (DE), Direct Pairs (DP), Cumulative Pairs (DP) and Naive Clustering (NC), have been tested against both real and synthetic instances. Results show the effectiveness of all methods but suggest that some structures work better than others. In particular, the simplest structure (DE) performs definitely better than the simple randomized restart MSH algorithm obtained by disabling the learning capability given by the pheromone and then using the Naive Clustering structure. DE is superior also to the CP method but no statistical evidence that DE is better than CE and DP has been found. Moreover, the behaviour of the algorithms seems to be sensitive to problem scaling.

Since there is a need, in robot skin design, for automated methods handling an ever increasing number of sensors per area unit, obtained thanks to miniaturization and technology improvements, even larger cases will be considered in future research activities.

Acknowledgement. The research leading to these results has received funding from the European Community's Seventh Framework Programme (FP7/2007-2013) under grant agreement no. 231500/ROBOSKIN.

References

1. Alirezarei, H., Nagakubo, A., Kuniyoshi, Y.: A highly stretchable tactile distribution sensor for smooth surfaced humanoids. In: Proceedings of the 2007 IEEE Conference on Humanoid Robotics (HUMANOIDS 2007), Pittsburg, PA, USA (2007)
2. Allen, P., Michelman, P.: Acquisition and interpretation of 3-D sensor data from touch. IEEE Transactions on Robotics and Automation 6(4), 397–405 (1990)
3. Anghinolfi, D., Cannata, G., Mastrogiovanni, F., Nattero, C., Paolucci, M.: Heuristic approaches for the optimal wiring in large-scale robotic skin design. Computers & Operations Research 39(11), 2715–2724 (2012)
4. Anghinolfi, D., Paolucci, M.: A new ant colony optimization approach for the single machine total weighted tardiness scheduling problem. International Journal of Operations Research 5, 1–17 (2008)
5. Argall, B., Billard, A.: A survey of tactile human-robot interactions. Robotics and Autonomous Systems 58(10), 1159–1176 (2010)
6. Baglini, E., Cannata, G., Mastrogiovanni, F.: Design of an embedded networking infrastructure for whole-body tactile sensing in humanoid robots. In: Proceedings of the 2010 IEEE Conference on Humanoid Robotics (HUMANOIDS 2010), Nashville, TN, USA (2010)
7. Bazgan, C., Couëtoux, B., Tuza, Z.: Complexity and approximation of the constrained forest problem. Theoretical Computer Science 412(32), 4081–4091 (2011)
8. Cannata, G., Dahiya, R., Maggiali, M., Mastrogiovanni, F., Metta, G., Valle, M.: Modular skin for humanoid robot systems. In: Proceedings of the 4th International Conference on Cognitive Systems (CogSys 2010), Zurich, Switzerland (2010)
9. Cannata, G., Denei, S., Mastrogiovanni, F.: A framework for representing interaction tasks based on tactile data. In: Proceedings of the 2010 IEEE International Symposium on Robot and Human Interactive Communication (RO-MAN 2010), Viareggio, Italy (2010)
10. Cannata, G., Denei, S., Mastrogiovanni, F.: Tactile sensing: Steps to artificial somatosensory maps. In: Proceedings of the 2010 IEEE International Symposium on Robot and Human Interactive Communication (RO-MAN 2010), Viareggio, Italy (2010)
11. Cannata, G., Denei, S., Mastrogiovanni, F.: Towards automated self-calibration of robot skin. In: Proceedings of the 2010 IEEE International Conference on Robotics and Automation (ICRA 2010), Anchorage, Alaska, USA (2010)
12. Cannata, G., Maggiali, M., Metta, G., Sandini, G.: An embedded artificial skin for humanoid robots. In: Proceedings of the 2008 IEEE International Conference on Multi-sensor Fusion and Integration (MFI 2008), Seoul, South Korea (2008)

13. Chang, M.C.W., Tsao, L., Yang, S., Yang, Y., Shih, W., Chang, F., Chang, S., Fan, K.: Design and fabrication of an artificial skin using pi-copper film. In: Proceedings of the 20th IEEE International Conference on Micro Electro Mechanical Systems (MEMS 2007), Kobe, Japan (2007)
14. Cordone, R., Maffioli, F.: Coloured Ant System and Local Search to Design Local Telecommunication Networks. In: Boers, E.J.W., Gottlieb, J., Lanzi, P.L., Smith, R.E., Cagnoni, S., Hart, E., Raidl, G.R., Tijink, H. (eds.) EvoWorkshop 2001. LNCS, vol. 2037, pp. 60–69. Springer, Heidelberg (2001)
15. Dahiya, R., Metta, G., Valle, M., Sandini, G.: Tactile sensing: from humans to humanoids. IEEE Transactions on Robotics 26(1), 1–20 (2010)
16. Denei, S., Mastrogiovanni, F., Cannata, G.: Parallel force-position control mediated by tactile maps for robot contact tasks. In: Proceedings of the 2012 IEEE International Conference on Intelligent Robots and Systems (IROS 2012), Vilamoura, Portugal (2012)
17. Dorigo, M., Blum, C.: Ant colony optimization theory: A survey. Theoretical Computer Science 344, 243–278 (2005)
18. Dorigo, M., Gambardella, L.: Ant colony system: a cooperative learning approach to the traveling salesman problem. IEEE Trans. on Evolutionary Computation 1(1), 53–66 (1997)
19. Duchaine, V., Lauzier, N., Baril, M., Lacasse, M., Gosselin, C.: A flexible robot skin for safe physical human robot interaction. In: Proceedings of the 2009 IEEE International Conference on Robotics and Automation (ICRA 2009), Kobe, Japan (2009)
20. Feder, T., Hell, P., Klein, S., Motwani, R.: Complexity of graph partition problems. In: Proceedings of the 1999 Annual ACM Symposium on Theory of Computing (STOC 1999), Atlanta, GA, USA (1999)
21. Futai, N., Yasuda, T., Inaba, M., Shimoiama, I., Inoue, H.: A soft tactile sensor with films of lc resonant traps. In: Proceedings of the 1999 IEEE International Conference on Advanced Robotics (ICAR 1999), Tokio, Japan (1999)
22. Goger, D., Worn, H.: A highly versatile and robust tactile sensing system. In: Proceedings of the 2007 IEEE Conference on Sensors (SENSORS 2007), Atlanta, GA, USA (2007)
23. Gupta, A., Lafferty, J., Liu, H., Wasserman, L., Xiu, M.: Forest density estimation, Tech. rep., Carnegie Mellon University (2010)
24. Hakozaki, M., Oasa, H., Shinoda, H.: Telemetric robot skin. In: Proceedings of the 1999 IEEE International Conference on Robotics and Automation (ICRA 1999), Detroit, Michigan, USA (1999)
25. Hakozaki, M., Shinoda, H.: Digital tactile sensing elements communicating through conducting skin layers. In: Proceedings of the 2002 IEEE International Conference on Robotics and Automation (ICRA 2002), Washington, DC (2002)
26. Hoshi, T., Shinoda, H.: A large area robot skin based on cell-bridge system. In: Proceedings of the 2006 IEEE Conference on Sensors (SENSORS 2006), Daegu, Korea (2006a)
27. Hoshi, T., Shinoda, H.: A sensitive skin based on touch-area-evaluating tactile elements. In: Proceedings of the 2006 IEEE Symposium on Haptic Interfaces for Virtual Environment and Teleoperator Systems, Alexandria, Virginia, USA (2006b)
28. Hwang, E., Seo, J., Kim, Y.: A polymer-based flexible tactile sensor for both normal and shear load detection and its application to robotics. Journal of Microelectromechanical Systems 16(3), 556–564 (2007)
29. Imielinska, C., Kalantari, B., Khachiyan, L.: A greedy heuristic for a minimum-weight forest problem. Operations Research Letters 14(2), 65–71 (1993)

30. Jockursh, J., Walter, J., Ritter, H.: A tactile sensor system for a three fingered robot manipulator. In: Proceedings of the 1997 IEEE International Conference on Robotics and Automation (ICRA 1997), Albuquerque, New Mexico, USA (1997)

31. Lacasse, M., Duchaine, V., Gosselin, C.: Characterization of the electrical resistance of carbon-black-filled silicone: Application to a flexible and stretchable robot skin. In: Proceedings of the 2010 IEEE International Conference on Robotics and Automation (ICRA 2010), Anchorage, Alaska (2010)

32. Lazzarini, R., Magni, R., Dario, P.: A tactile array sensor layered in an artificial skin. In: Proceedings of the 1995 IEEE/RSJ International Conference on Intelligent Robots and Systems (IROS 1995), Pittsburg, PA, USA (1995)

33. Miller, C.E., Tucker, A.W., Zemlin, R.A.: Integer programming formulation of traveling salesman problems. J. ACM 7, 326–329 (1960)

34. Mittendorfer, P., Cheng, G.: Humanoid multimodal tactile-sensing modules. IEEE Transactions on Robotics 27(3), 401–410 (2011)

35. Monnot, J., Toulouse, S.: The path partition problem and related problems in bipartite graphs. Operations Research Letters 35(5), 677–684 (2007)

36. Nagakubo, A., Alirezarei, H., Kuniyoshi, Y.: A deformable and deformation sensitive tactile distribution sensor. In: Proceedings of the 2007 IEEE International Conference on Robotics and Biomimetics (ROBIO 2007), Sanya, China (2007)

37. Nattero, C., Anghinolfi, D., Paolucci, M., Cannata, G., Mastrogiovanni, F.: Experimental analysis of different pheromone structures in ant colony optimization for robotic skin design. In: Proceedings of the 2012 International Federated Conference on Computer Science and Information Systems (FedCSIS 2012), Wrocland, Poland (2012)

38. Nilsson, M.: Tactile sensing with minimal wiring complexity. In: Proceedings of the 1999 IEEE International Conference on Robotics and Automation (ICRA 1999), Detroit, Michigan, USA (1999)

39. Papadimitriou, C., Steiglitz, K.: Combinatorial Optimization: Algorithms and Complexity. Prentice-Hall, Upper Saddle River (1982)

40. Schaeffer, S.: Graph clustering. Cluster Science Review 1, 27–64 (2007)

41. Schmitz, A., Maiolino, P., Maggiali, M., Natale, L., Cannata, G., Metta, G.: Methods and technologies for the implementation of large-scale robot tactile sensors. IEEE Transactions on Robotics 27(3), 389–400 (2011)

42. Shimojo, M.: Spatial filtering characteristic of elastic cover for tactile sensor. In: Proceedings of the 1994 IEEE International Conference on Robotics and Automation (ICRA1993), San Diego, CA, USA (1994)

43. Shimojo, M., Araki, T., Ming, A., Ishikawa, M.: A high speed mesh of tactile sensors fitting arbitrary surfaces. IEEE Sensors Journal 10(4), 822–831 (2010)

44. Shinoda, H., Oasa, H.: Wireless tactile sensing element using stress-sensitive resonator. IEEE/ASME Transactions on Mechatronics 5(3), 258–266 (2000)

45. Stützle, T., Hoos, H.: Max-min ant system. Future Generation Computer Systems 16, 889–914 (2000)

46. Yan, X., Zhou, X., Han, J.: Mining closed relational graphs with connectivity constraints. In: Proceedings of the 2005 International Conference on Knowledge Discovery and Data Mining (KDD 2005), Chicago, Illinois, USA (2005)

47. Youssefi, S., Denei, S., Mastrogiovanni, F., Cannata, G.: A middleware for whole body skin-like tactile systems. In: Proceedings of the 2011 IEEE-RAS International Conference on Humanoid Robotics (Humanoids 2011), Bled, Slovenia (2011)

Homogeneous Non Idling Problems:
Models and Algorithms

Alain Quilliot[1], Philippe Chretienne[2], and Benoit Bernay[1]

[1] LIMOS, UMR CNRS 6158, Bat. ISIMA, Universit BLAISE PASCAL,
Campus des Czeaux, BP 125, 63173 AUBIERE, France
`alain.quilliot@isima.fr`
[2] LIP6, Universit PARIS VI, Place JUSSIEU, 75005, PARIS, France

Abstract. This paper is about multi-processor scheduling with non idling constraints, *i.e.* constraints which forbid interruption in the use of the processors. We first reformulate the problem, while using a notion of pyramidal shape, and next apply matching techniques in order to get a min-max feasibility characterization criterion, which allows us to derive a polynomial algorithm for the related existence problem and for the Makespan Minimization related problem. The last part of the paper is devoted to the Linear Cost Minimization of multiprocessor scheduling with non idling constraints, which we handle through linear model and a heuristic Lagrangean decomposition approach, and involves numerical experiments.

Keywords: Multiprocessor Scheduling, Matching Theory.

1 Introduction

Idling scheduling means that a machine waits between the completion of a job and the start of the next job. Moreover, it is well-known that such waiting delays are often necessary to get optimality, whatever be the related performance criterion. This is the key feature which explains why list algorithms, which do not allow a machine to wait for a more urgent job, do not generally yield optimal schedules. However, it may happen that in some applications such that those described in [5], the cost of making a running machine stop and restart later is so high that a non idling constraint is put on the machine so that only schedules without any intermediate delays are accepted. For instance, if the machine is an oven which must heat different and non compatible pieces of work at a given high temperature, keeping the required temperature of the oven while it is empty may clearly become too costly. Problems concerning power management policies may also yields similar constraints [6], where for example each idling period has a cost and the total cost has to be minimized [1]. Note that the non idling constraint will not necessarily ensure full machine utilization but will remove the cost of machine re-starts, maybe at the price of processing the jobs later.

S. Fidanova (Ed.): *Recent Advances in Computational Optimization*, SCI 470, pp. 115–134.
DOI: 10.1007/978-3-319-00410-5_7   © Springer International Publishing Switzerland 2013

2   Notations, Problem Definition and Reformulation

2.1   Main Notations: Time-Units and Intervals

We consider the discrete time space $N = \{0, .., +\}$, each element being a *time-unit*. A finite subset Ω of N is an *interval* if it is made of consecutive time-units. The smallest (largest) time-unit of an interval Ω is denoted $min(\Omega)(max(\Omega))$. If p and q are two distinct natural numbers, the interval whose bounds are p and q is denoted by $I(p, q)$ (note that we may have $p < q$ or $q < p$). Interval Ω_2 is said to dominate interval Ω_1 if $max(\Omega_1) + 1 < min(\Omega_2)$: such a situation is denoted by $\Omega_1 Dom \ \Omega_2$. If $\Omega_1 Dom \ \Omega_2$, then we denote by $Mid(\Omega_1, \Omega_2)$ the (non empty) interval $max(\Omega_1) + 1, .., min(\Omega_2) - 1$. We say that two intervals and are connected if their union is an interval.

2.2   Main Notations: Time-Units and Intervals

We now suppose that we are given a set $J = \{J_1, .., J_n\}$ of n unit-time jobs that are to be processed on a set $M = \{M_1, .., M_m\}$ of m identical machines. Job J_i must be executed inside a given time-window $F(i) = \{r_i, .., d_i\}$ which is an interval. It will be convenient to denote r_{min} (respectively d_{max}) the smallest r_i (respectively the smallest d_i) and by H the interval $\{r_{min}, .., d_{max}\}$. The jobs are also constrained by a *weak* precedence relation denoted by $<<$ where $J_i << J_j$ means that J_j must not be performed before J_i. Jobs are further constrained by the so-called homogeneous non-idling constraint (*HNI* in short) which imposes that, for any subset M of M, the time units at which the machines of M are busy define an interval. Then a schedule of the job set M is a pair (T, μ), where T and μ are two functions, which assign, to any job J_i, respectively a time-unit $T(i)$ and a machine $\mu(i)$. The schedule (T, m) is said to be *feasible* if:

- For any job $J_i, T(i) \in F(i)$;
- For any pair of jobs J_i, J_j, such that $J_i << J_i$, we have $T(i)T(j)$;
- For any pair of jobs J_i, J_j, we have either $T(i)T(j)$ or $\mu(i)\mu(j)$;
- The *HNI* condition is satisfied: for any subset M of M, the set $t \in N$, such that there exist $i = 1..n$, with $T(i) = t$ and $T(i) \in M$ is an interval.

For any such a schedule (T, μ), we define the *active time-unit* set of the function T as the image set of T, that means as the subset $ACT(T)$ of $\mathbf{N}$ defined by: $ACT(T) = \{t \in N$ such that there exists at least some index value $i = 1..n$, with $T(i) = t\}$.

Then we define the related *Feasibility Homogeneous Non-Idling Scheduling Problem* NON-IDLE$_0 = (P, HNI|p_i = 1, r_i, d_i, preceq|-)$ as the problem which consists in deciding whether the given instance $(J, F, <<, m)$ admits at least one feasible schedule. If the answer is yes, we say that this instance is *feasible*.

It must be pointed out that the precedence relation *preceq* which we handle here has not the same meaning as the classical one, since when we set $J_i << J_i$, we allow J_i and J_j to be processed at the same time-unit. Clearly, due to the machine constraint, there is no difference when $m = 1$.

It is also of interest to notice that the problem $(P, HNI|p_i = 1, r_i, d_i, preceq|-)$, where $prec$ is the usual precedence relation, is NP-Complete, since the NP-Complete $(P|p_i = 1, prec|C_{max}d)$, polynomially reduces to the problem $(P, HNI|p_i = 1, r_i, d_i, preceq|-)$, by adding (mdn) filling jobs and setting $r_i = 1$ and $d_i = d$ for all the jobs.

2.3 A Reformulation through Pyramidal Shape Functions

Let $I = (J, F, <<, m)$ be an instance of NON-IDLE$_0$ and (T, μ) some schedule of I. If $t \in \mathbf{N}$, we denote by $n_T(t) = Card(T^{-1}(t))$ the number of jobs which are scheduled at time-unit t according to T, and we call this function the *resource profile function* of the schedule. We say that this function $t \to n_T(t)$ has a pyramidal shape if for any time units t, t, t such that $t < t < t$, we have $Inf(n_T(t), n_T(t))n_T(t)$ (see fig. 1).

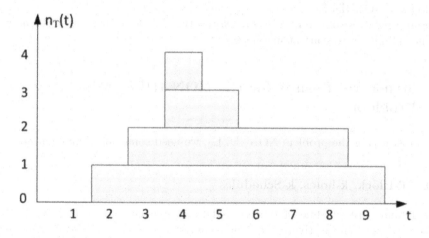

Fig. 1. A pyramidal function $n_T(t)$

Then we say that (T, μ) is a *flat schedule* of the instance I, if for any time unit t such that $n_T(t) > 0$, we have $\mu(T^{-1}) = \{M_1, .., M_{n_T(t)}\}$. Clearly any feasible schedule (T, μ) can be turned, through reassignment of the jobs onto the machines, into a flat feasible schedule, and, in the case of a flat schedule, the knowledge of T determines μ. So, the following statement reformulates the NON-IDLE$_0$ as a problem which only involves the time function T, subject to some *pyramidal shape* property related to the profile function $t \to n_T(t)$.

Theorem 1: *Solving the NON-IDLE$_0$ problem in the case of instance $I = (J, F, <<, m)$ only means computing the function T, which with any job J_i associates some time-unit $T(i)$ in such a way that:*

1. *For any job $J_i, T(i) \in F(i)$;*

2. *For any pair of jobs J_i, J_j, such that $J_i << J_j$, we have $T(i)T(j)$;*
3. *For any time-unit t, $n_T(t) = Card(T^{-1}(t))m = $ the number of machines;*
4. *The function $t \to n_T(t)$ has a pyramidal shape.*

A function T which satisfies 1 and 3 above is called a *m-matching*. In case it also satisfies 2, it is called a *pre-schedule*. In case it satisfies all conditions 1..4, we shall also call T a *feasible schedule* of the NON-IDLE$_0$ of the instance $I = (J, F, <<, m)$.

2.4   Makespan Minimization Homogenous Non-idling Scheduling Problem

Let $I = (J, F, <<, m)$ as above, and let T be some feasible schedule for I. The Makespan of T is the cardinality of its active time-unit set $ACT(T)$. Then the *Makespan Minimization Homogeneous Non-Idling Problem* NON-IDLE$_1$ comes as follows: NON-IDLE$_1$: *Compute a feasible schedule T of $I = (J, F, <<, m)$ with a minimal makespan $Card(ACT(T))$.* Notice that, while setting this problem, we do not require T to start at instant 0.

3   Structural Results for the NON-IDLE$_0$ Feasibility Problem

Before studying the problem NON-IDLE$_0$, we need some additional concepts.

3.1   T-Block, k-holes, k-Schedules

Let T some pre-schedule of the NON-IDLE$_0$ instance $I = (J, F, <<, m)$. We denote by $s(T)$ (respectively $e(T)$) the smallest (respectively largest) time-unit such that at least one job is performed at time-unit t. We say that an interval $\Omega \subseteq ACT(T) = \{s(T), .., e(T)\}$ is a $T - block$ if every job J_i which is scheduled inside Ω is such that $F(i) \subseteq \Omega$. A time-unit t is then a $k - hole$ for T, where k is some positive number, if there exists time-units t, t, t such that:

- $t < t < t$;
- $Inf(n_T(t), n_T(t)) > n_T(t) = k$ (in the following figure 2, timeunit 6 is a 2-hole and time-unit 7 is a 1-hole).

Clearly, T is a feasible schedule if only if, for any k in $\{0..m - 1\}$, it has no $k - hole$. This leads us to introduce the intermediate notion of $k - schedule$: the pre-schedule T is a $k - schedule$ if T has no $l - hole$ for $l = 0..k - 1$, or, equivalently, if the function $t \to Inf(k, n_T(t))$ has a pyramidal shape. The time diagram of Figure 2 represents a pre-schedule which is a 1-schedule, but not a 2-schedule since time-unit 7 is a 1-hole. Obviously, a feasible schedule is a $m - schedule$ and conversely.

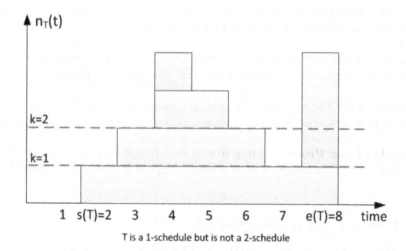

T is a 1-schedule but is not a 2-schedule

Fig. 2. k-holes

3.2   Time-Windows Stability

The family of time-windows $F(i), i = 1..n$, is *stable* with respect to the precedence relation $<<$ if, for any pair of jobs J_i, J_j such that $J_i << J_j$, we have: $r_i r_j$ and $d_i d_j$. The following result, which is almost obvious, shows that we may assume, without any loss of generality, that our **input family of time windows is stable with respect to the precedence relation** $<<$:

Proposition 1:   *Let* $I = (J, F, <<, m)$ *be an instance of the NON-IDLE$_0$ problem. There exists an instance* $I = (J, F, <<, m)$, *which may be obtained from* I *through constraint propagation, admits the same set of feasible solution as* f, *and is such that: F is stable with respect to* $<<$ *and, for any I,* $F(i) \subseteq F(i)$.

Since the goal of this paper is mainly to provide a characterization of the feasible instances of the NON-IDLE$_0$ problem, together with recognition and makespan minimization algorithms, we proceed in several steps. First, we use the Konig-Hall Theorem in order to characterize the instances which admits a $m-matching$. Then, we show that any $m - matching$ may be turned into a pre-schedule. We keep on by identifying a structural property which is going to make possible turning this pre-schedule into a feasible schedule. Finally, we translate this mathematical characterization into a recognition algorithm.

3.3   Existence of a m-Matching and of Pre-schedule

Let $I = (J, F, <<, m)$ some NON-IDLE$_0$ instance. For any interval Ω, we denote by $J(\Omega)$ the set of all jobs J_i, such that $F(i) \subseteq \Omega$. Then one may derive in a straightforward way from classical Konig-Hall Theorem related to the existence of generalized matching in bipartite graphs that:

Proposition 2: *The instance $(J, F, <<, m)$ admits a $m - matching$ if and only if, for any interval Ω of N, we have $Card(J(\Omega)) m.Card(\Omega)$.*

The next property is mainly due to our assumption about the stability of F. It will be essential for us when we deal with the precedence relation $<<$.

Proposition 3: *The instance $(J, F, <<, m)$ admits a pre-schedule if an only if it admits a $m - matching$.*

Principle of the Proof: starting from a $m - matchingT$, one iteratively exchanges $T(i), T(j)$ values in order to make T compatible with the $<<$ precedence relation.

3.4   Existence of a Feasible Schedule

Let $I = (J, F, <<, m)$ our NON-IDLE$_0$ instance. If the condition provided by Proposition 3 is satisfied, we easily become able to produce some pre-schedule T. However, such a pre-schedule may have k-holes and thus may not fit the pyramidal property required for the feasible schedules. This section will provide an additional necessary and sufficient condition so that an instance $I = (J, F, << , m)$ admits some feasible schedule. Before deriving this condition, we first give two simple lower bound properties which must be met by such a feasible schedule and then introduce the notion of propagation path which will prove to be a quite useful tool either to transform a pre-schedule into a feasible schedule or to prove the no existence of such a feasible schedule.

Let Ω be an interval of N. We denote by $Int(\Omega)$ the set of intervals which are contained into Ω and by $\lambda(\Omega)$ the integer value $Sup_{\omega \in Int(\Omega)} \lceil Card(J(\omega))/Card(\omega) \rceil$. Then the meaning of those definitions comes in an immediate way through the almost trivial following lemma:

Lemma 1: *Let T be a $m - matching$ of $I = (J, F, <<, m)$ and let Ω be an interval of N. There must exist at least one time-unit in Ω such that at least $\lambda(\Omega)$ machines are busy.*

We understand the main role of $\lambda(\Omega)$ is to provide us with a lower bound of the number of machines which are going to be necessary if we want to succeed in scheduling the jobs of $J(\Omega)$. Notice that if Ω is a $T - block$, then we have $\lambda(\Omega) \lceil \sum_{t \in \Omega} n_T(t)/Card(\Omega) \rceil$, since, in this case, $J(\Omega)$ is exactly the set of the jobs which are scheduled inside Ω.

Let us assume now that Ω_1 and Ω_2 are two intervals of N such that $\Omega_1 Dom \Omega_2$. If we denote by $\mu(\Omega_1, \Omega_2)$ the value $Card(Mid(\Omega_1, \Omega_2)).Inf(\lambda(\Omega_1), \lambda(\Omega_2))$, then we easily see that we get:

Lemma 2: *In any feasible schedule of $I = (J, F, <<, m)$, at least $\mu(\Omega_1, \Omega_2)$ jobs are scheduled in the interval $Mid(\Omega_1, \Omega_2)$.*

Also, the following two properties, which are related to $T - blocks$, will be useful in order to derive the main characterization result:

Lemma 3:   *Let T be some $m - matching$ of $I = (J, F, <<, m)$ and let Ω_1 and Ω_2 be two connected $T - blocks$. Then $\Omega_1 \cup \Omega_2$ is a $T - block$ and $\lambda(\Omega_1 \cup \Omega_2)Sup(\lambda(\Omega_1), \lambda(\Omega_2))$.*

Lemma 4:   *Let T be a $m-matching$ of $I = (J, F, <<, m)$, let Ω_1 be a $T-block$, and let Ω_1 be an interval such that $\Omega_1 \cap \Omega_2$ is empty. If, for any pair (u, t) in $\Omega_1 * \Omega_2, n_T(u)n_T(t)$ (respectively $n_T(u) > n_T(t)$), then $\lambda(\Omega_1)\lambda(\Omega_2)$ (respectively $\lambda(\Omega_1) > \lambda(\Omega_2)$).*

Given a $m - matching$ T, we now define what is a *propagation path* of T. We first define *the propagation graph* $G = (H, E(T))$ as the labeled directed graph whose node set is the interval $H = r_{min}, .., r_{max}$ of the possible values for T, and the arc set $E(T)$ is defined by:

- $[t, t] \in E(T))$ if tt and there is at least one job J_i which is scheduled at t and which is such that $t \in F(i)$.
- If job J_i is scheduled at t and $t \in F(i)$, then J_i is said to be a *label* of the arc $[t, t]$.

Then, a *propagation path* of T is an elementary path $\gamma = (t_0, .., t_k)$ of $G(T)$. The subpath of γ from t_i to t_j will be denoted by $\gamma(t_i, t_j)$. The *length* $L(\gamma)$ of γ is the value $k + 1$ (that means the number of vertices of γ), and the *extended length* $L^*(\gamma)$ of γ is the sum $\sum_{s=1..k} |t_s t_{s-1}|$.

This propagation path γ is *monotone* if the sequence $(t_0, .., t_k)$ is either decreasing or increasing. It is $no - cross$ if for any $r \in 1..k$, we have either $t_r > Sup_{i=1..r-1} t_i$ or $t_r < Inf_{i=1..r-1} t_i$. Clearly, the no-cross property may be viewed as a weak version of monotonicity. It is *labeled* when all its arcs are assigned with labels, that means when, with any arc in γ, we decided to associate some job J_i whose value $T(i)$ is likely to be modified through shift propagation along γ.

The path γ is said to be *fitted* if it is labeled and satisfies $Card(T^{-1}(t_k)) < m$. Of course, we understand that if $\gamma = (t_0, .., t_k)$ is a propagation path of $G(T)$, which is fitted and provided with the labeling $\sigma = (J_i(0), .., J_i(k - 1))$, then we become able to modify the $m - matching$ T and get another $m - matching$ $T = Trans(T, \gamma, \sigma)$ by setting $T(J_i(p)) = t_{p+1}$ for any $p = 0, .., k - 1$.

Let T be a pre-schedule and let $\gamma = (t_0, .., t_k)$ be a propagation path of the graph $G(T)$. If γ has at least one labeling σ such that $T = Trans(T, \gamma, \sigma)$ is a preschedule, then γ is *compatible* with the precedence relation $<<$ ($<< -compatible$ in short).

The two following lemmas are fundamental tools, which show that no-cross propagation paths allow to transform a pre-schedule into another one:

Lemma 5:   *Let T be a pre-schedule and let us assume that $\gamma = (t_0, .., t_k)$ is a propagation path of $G(T)$ from $t_0 = u$ to $t_k = v$. Then there exists a no-cross propagation path from u to v in $G(T)$.*

Proof: Assume that γ is not no-cross and let $t_r (2rq)$ be the first node of γ such that $min_{i=1..r-1} t_i < t_r < max_{i=1..r-1} t_i$. From the definition of t_r, we know that there is a smallest index $s(0sr-2)$ such that t_r belongs to $I(t_s, t_{s+1})$. Thus $[t_s, t_r]$ is an arc of $G(T)$ and the concatenation $\gamma(t_0, t_s).[t_s, t_r]$ is no-cross. The above transformation may then be iterated while the current propagation path is not no-cross.

Lemma 6:   *Let T be a pre-schedule and let us assume that $\gamma = (t_0, .., t_k)$ is a propagation path of $G(T)$ from $t_0 = u$ to $t_k = v$, where $Card(T^{-1}(t_k)) < m$. Then there exists a no-cross and $<< -compatible$ propagation path from u to v in $G(T)$.*

Sketch of the Proof. Define the extended length of such a path γ in G(T) as the sum $\sum_{i=1..k} |t_i t_{i-1}|$, and consider a propagation path γ from u to v with minimal extended length and whose length is maximal among the paths with minimal extended length. Assume also that γ is not $<< -compatible$ and let $\sigma = (J_{i(0)}, .., J_{i(k-1)})$ be a labeling of γ. The jobs of the labeling are called the *moving* jobs, while the other jobs are called the *static* jobs. Since T is a pre-schedule and γ is not $<< -compatible$, $<<$ is violated in $T = Trans(T, \gamma, \sigma)$ either because an inversion has been created between a static job J scheduled at t and a moving job $J_{i(s)}$ scheduled at t_{s+1} or between two moving jobs $J_{i(s)}$ and $J_{i(r)}$ respectively scheduled at time t_{s+1} and t_{r+1}. In both case, one checks that it is possible to rearrange path γ in order to get a contradiction on the minimality of γ.

Let T be a pre-schedule and let u be a time-unit such that $n_T(u) > 0$. The next lemma, whose proof is essentially routine, provides us with an important property of the set $A_T(u)$ of the time-units that may be reached from u by the propagation paths of $G(T)$.

Lemma 7: *The set $A_T(u)$ is a T-block.*

We are now able to describe and state the structural condition which must be met by an instance $I = (J, F, <<, m)$ of the NON-IDLE$_0$ problem so that it admits some feasible schedule.

Theorem 2:   *The instance $I = (J, F, <<, m)$ of NON-IDLE$_0$ is feasible if and only if the following two conditions are satisfied:*

1. *For any interval Ω of N, $Card(J(\Omega))m.Card(\Omega)$.*
2. *For any sequence $(\Omega_1, ..\Omega_p)$ of intervals of N, such that $\Omega_1 DomDom\Omega_p$, we have: $Card(J - \cup_{s=1..p} J(\Omega_s)) \sum_{s=1..p-1} \mu(\Omega_s, \Omega_{s+1})$.*

Sketch of the Proof. The only if part of the proof is a straightforward consequence of propositions 2 and 4 and lemmas 1, 2, 3

Before starting the proof of the part if, let us recall that, for any $k = 1..m$, a pre-schedule T is a $k - schedule$ if T has no $l - hole$ for any $l = 0..k - 1$. So we may adapt the definition of the quantity $\mu(\Omega_1, \Omega_2)$, where Ω_1 and Ω_2 are two time intervals such that Ω_1 Dom Ω_2, to $k - schedules$ by setting: $\mu^*(\Omega_1, \Omega_2, k) = Card(Mid(\Omega_1, \Omega_2)).Inf(\lambda(\Omega_1), \lambda(\Omega_2), k)$. Then one easily checks:

Lemma 8: *Let Ω_1 and Ω_2 be two time intervals such that Ω_1 Dom Ω_2. In any k-schedule of $I = (J, F, <<, m)$, at least $\mu^*(\Omega_1, \Omega_2, k)$ jobs are scheduled in the interval $Mid(\Omega_1, \Omega_2)$.*

In order to prove the if part, we extend the statement of Theorem 2 to $k - schedules$ and prove it by induction on k. Using $k - schedules$ allows us to perform an inductive reasoning in order to prove Theorem 2. As a matter of fact, what we have to do is to prove the following statement, which may be viewed as an inductive extension of Theorem 2 to $k - schedules$:

Inductive Formulation of Theorem 2: *The instance $I = (J, F, <<, m)$ of NON-IDLE$_0$ admits at least one $k - schedule$ if and only if the following two conditions are satisfied:*

1. *For any interval Ω of N, $Card(J(\Omega))m.Card(\Omega)$.*
2. *For any sequence $(\Omega_1, ..\Omega_p)$ of intervals of N, such that $\Omega_1 Dom Dom\ \Omega_p$, we have: $Card(J - \cup_{s=1..p}J(\Omega_s)) \sum_{s=1..p-1} \mu^*(\Omega_s, \Omega_{s+1}, k)$.*

In order to prove this last statement, we proceed by induction on k. More precisely, we show that if an instance $I = (J, F, <<, m)$ of the NON-IDLE$_0$ problem has at least a $k - schedule$ and no $(k+1) - schedule$, then there exists a sequence of intervals that does not satisfy condition 2 of the above statement. In order to get such a sequence, we consider a $k - schedule$ of I, which is doubly minimum in the following sense:

- T has a minimum number of $k - holes$;
- the vector $(N_T(m), .., N_T(1))$, where $N_T(j)$ is the number of time-units t such that $n_T(t) = j$, is lexicographically minimum.

Thus, T may be viewed as the flattest $k - schedule$ with a minimum number of $k - holes$.

The graph of the piecewise constant function $t \rightarrow n_T(t)$ may be decomposed into:

- a *left* part, which is a *climbing stair* (increasing piecewise constant function) for which we denote by $C_p^{Inf}, 1pk + 1$, the smallest time-unit at which the stair height is at least equal to p;
- a *right* part, which is a *descending stair* (piecewise constant decreasing function) for which we denote by $C_p^{Sup}, 1pk + 1$, the largest time-unit at which the stair height is at least equal to p;
- a *medium* part, which is made of the time-units u such that $n_T(u)k + 1$, and which is such that at least one time-unit is a $k - hole$.

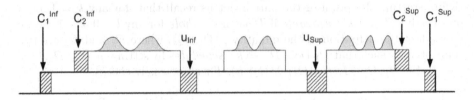

Fig. 3. $t \to n_T(t)$ segmentation

From the minimality of T, we easily get that the propagation graph $G(T)$ is such that there is no propagation path:

- from t which satisfies $n_T(u)k + 2$ to u which is a $k - hole$;   (E1)
- from a time-unit C_p^{Sup} or C_p^{Inf} to a time-unit u which is a $k - hole$;   (E2)
- from a time-unit C_p^{Sup} to a time-unit $C_j^{Sup} + 1, 1j < p$;   (E3)
- from a time-unit C_p^{Inf} to a time-unit $C_j^{Inf} + 1, 1j < p$;   (E4)

From what precedes, we deduce 3 families of intervals:

- the *right* family $L_1..L_l$, which we get from the $T - blocks$ $A_T(C_p^{Sup}), 1pk+1$, by merging those intervals which are connected. By using lemmas 7 and 8, we may check that, for any $i = 1..l - 1$: $\sum_{t \in Mid(L_i,L_{i+1})} n_T(t) = \lambda(L_i).Card(Mid(L_i, L_{i+1}))\mu^*(L_i, L_{i+1}, k + 1)$;
- the *left* family $L_1^*..L_h^*$, which we get from the $T - blocks$ $A_T(C_p^{Inf}), 1pk+1$, by merging those intervals which are connected. By using lemmas 7 and 8, we may check that, for any $i = 1..h - 1$: $\sum_{t \in Mid(L_i^*,L_{i+1}^*)} n_T(t) = \lambda(L_i^*).Card(Mid(L_i^*, L_{i+1}^*))\mu^*(L_i^*, L_{i+1}^*, k + 1)$;
- the *medium* family $J_1..J_q$, which we get from the $T - blocks$ $A_T(u)$, u such that $n_T(u)k + 2$, by merging those intervals which are connected. By using lemmas 7 and 8, we check that, for any $i = 1..l - 1$: $\sum_{t \in Mid(J_i,J_{i+1})} n_T(t) = \lambda(J_i).Card(Mid(J_i, J_{i+1}))\mu^*(J_i, J_i + 1, k + 1)$;

The key point is that, because of (E1, .., E4), those intervals define several distinct connected blocks, with non empty space between them (we must find at

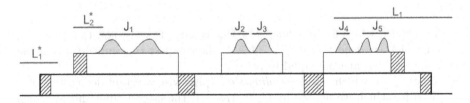

Fig. 4. $t \to n_T(t)$ Decomposition

least the k-holes, which are going to separate those intervals. Still, since we may have $L_1^* \cap J_1 Nil$ or $L_1 \cap J_q Nil$, we merge once again the connected intervals of those three families. Then we get a sequence of non connected intervals $M_1..M_r$, of disjoint $T - blocks$ such that $M_1 Dom..Dom M_r$ and that the number of jobs which is to in $\cup_{j=1..r} J(M_j)$ is not large enough to avoid the existence of a $k-hole$. Then we conclude.

4 Polynomial Algorithms for NON-IDLE$_0$ and NON-IDLE$_1$

4.1 A Polynomial Algorithm for the NON-IDLE$_0$ Feasibility Problem

The proof of Theorem 2 is not an algorithmic proof, since it involves an hypothesis about *doubly minimality* which has no algorithmic interpretation. In this section, we first show that a forbidden pattern of intervals may be derived from any $k - schedule$ which satisfies a weaker set of conditions than that of the doubly minimal $k - schedule$ T considered in the proof of Theorem 2. Its allow us to turn this minimality condition into conditions which might be used as halting test inside the while loop of a recog-nition algorithm. Then we derive from this a polynomial algorithm which solves NON-IDLE$_0$, and whose correctness mainly relies on this set of sufficient conditions.

So, let T be a $k - schedule$ and let U and V be two disjoint sets of time-units: we denote by $PP(U, V)$ the set of propagation paths of $G(T)$ which start in U and end into V. We denote by $Hole(k)$ the set of time-units t which are $k-holes$, and by $Top(k)$ the set of time-units t which satisfies $n_T(u)k + 2$. Then we may state:

Theorem 3: *Let $I = (J, F, <<, m)$ be an instance of the NON-IDLE$_0$ problem, and let k be some number in $1, .., m - 1$. Let us assume that T is a $k - schedule$ of I such that the following conditions are satisfied:*

1. *T is not a $(k + 1) - schedule$;*
2. *$PP(Top(k) \cup \{C_1^{Inf}, .., C_{k+1}^{Inf}\} \cup \{C_1^{Sup}, .., C_{k+1}^{Sup}\}), Hole(k))$ is empty;*
3. *$PP(Top(k), \{C_1^{Inf} - 1, .., C_{k+1}^{Inf} - 1\} \cup \{C_1^{Sup} + 1, .., C_{k+1}^{Sup} + 1\})$ is empty;*
4. *For $j = 2..k+1$, $PP(C_j^{Sup}, C_{j-1}^{Sup} + 1)$ is empty;*
5. *For $j = 2..k+1$, $PP(C_j^{Inf}, C_{j-1}^{Inf} - 1)$ is empty;*

Then the instance I has no (k+1)-schedule.

This result yields following algorithm SEARCH-SCHEDULE$(J, F, <<, m)$ which solves the NON-IDLE$_0$ problem by providing, for any instance $I = (J, F, << , m)$ either a feasible schedule of I or a $k - schedule$ T, (with $0km - 1$) of I

which satisfies the conditions 1..5 of the above statement. Notice that $Hole(k)$, $Top(k), C_p^{Sup}, C_p^{Inf}$ denote here variables that contains the values associated with the current $k - schedule\,T$.

Algorithm SEARCH-SCHEDULE $(J, F, <<, m)$:

$T \leftarrow Matching(J, F, m)$;
(*Computation of an initial m-matching,
through a standard matching procedure*)
If T does not exist (*Propositions 2 and 3*) then
      $SEARCH - SCHEDULE \leftarrow Fail$
Else
      $T \leftarrow Pre - Schedule(J, F, m, <<, T)$;
      (*Turn T into a pre-schedule through a
      sequence of exchanges, according to proposition 3*)
      $k \leftarrow Sup_{l=0..m}l$ such that T is a $l - schedule$; Not Stop;
      While $k < m$ and Not Stop do
            Try, according to this order, the existence of a propagation path γ in:
                $- PP(Top(k) \cup \{C_1^{Inf}, .., C_{k+1}^{Inf}\} \cup \{C_1^{Sup}, .., C_{k+1}^{Sup}\}, Hole(k))$;
                $- PP(Top(k), \{C_1^{Sup} + 1, .., C_{k+1}^{Sup} + 1\} \cup \{C_1^{Inf} - 1, .., C_{k+1}^{Inf} - 1\})$;
                $- \cup_{j=2..k+1} PP(\{C_j^{Sup}, C_j^{Inf}\}, \{C_{j-1}^{Inf} - 1, C_{j-1}^{Sup} + 1\})$;
            If Failure(Try) (*Non existence of γ*) then $Stop$ Else
                Let σ be a label of γ; $T \leftarrow Trans(T, \gamma, \sigma)$;
      If $k = m$ then $SEARCH - SCHEDULE \leftarrow T$
      else $SEARCH - SCHEDULE \leftarrow Fail$;

Theorem 4: *The above algorithm SEARCH-SCHEDULE solves the feasibility NON-IDLE$_0$ problem in polynomial time.*

Sketch of the Proof: one follows the proof of Theorem 2, while replacing the doubly minimality assumption by the conditions 1..5 of Theorem 3. As for complexity, it is easy to check that the above algorithm is time polynomial if the encoding size of a time-window $F(i)$ is defined as being proportional to the size of this time windows. But, in fact, it has to be related to the encoding size of both numbers $min(F(i))$ and $max(F(i))$. Still, one easily checks that, once an initial $m - matching$ has been computed, time-windows may be restricted in such a way that the size of their union be polynomially bounded by the number n of jobs, and that provides us with the key argument for the time-polynomiality of our algorithm.

4.2 A Polynomial Algorithm for the NON-IDLE$_1$ Problem

Makespan minimization is contained into feasibility testing, and comes in a simple way through the following process:

Makespan-Min-No-Idle-Schedule Algorithm.

Input: the instance $I = (J, F, <<, m)$
Output: a no idle feasible schedule or a Failure signal;
Initialize T through the SEARCH-SCHEDULE procedure;
If *Failure*(Initialize) then *Failure*
Else

> Not Stop;
> While Not Stop do
> $\Delta \leftarrow Makespan(T)$;
> Let t_1 and t_2 respectively the smallest
> and largest active time-units according to T;
> For any job $J_i \in J$, set $F_\Delta(i) = F(i) \cap \{t_1 + 1, t_2\}$;
> $\quad T - Aux \leftarrow SEARCH - SCHEDULE(J, F_\Delta, <<, m)$;
> $\quad$ If $T - AuxFailure$ then $T \leftarrow T - Aux$
> $\quad$ Else
> $\qquad$ For any job $J_i \in J$, set $F_\Delta(i) = F(i) \cap \{t_1, .., t_2 - 1\}$;
> $\qquad T - Aux \leftarrow SEARCH - SCHEDULE(J, F_\Delta, <<, m)$;
> $\qquad$ If $T - AuxFailure$ then $T \leftarrow T - Aux$ Else *Stop*.
> *Makespan-Min-No-Idle-Schedule* $\leftarrow T$;

Theorem 5: *The above Makespan-Min-No-Idle-Schedule algorithm solves the Makepan Minimization NON-IDLE$_1$ Problem in Polynomial Time.*

Sketch of the Proof: the basic point here is that if T is some feasible schedule with active time-unit set $ACT(T) = [a, b]$, and if there exists a feasible schedule T with smaller makespan than T, then T may be computed inside the time-window $[a, b]$.

5 Minimal Cost Homogeneous Non Idling Scheduling Problem

An other natural optimization formulation of the previously studied $NON - IDLE_0$ Problem comes from the hypothesis that the performance criterion involves specific running costs, which depend on both jobs and time units. In order to deal with this problem, we suppose that we are provided with an instance $I = (J, F, <<, m)$ of the NON-IDLE$_0$ problem, that the time space $\mathbf{N}$ is described in a explicit way as a finite discrete set $\mathbf{N} = 1..TS$, and that we are also provided with costs $C_{i,t}, i = 1..n, t \in \mathbf{N}$, whose meaning is: performing job J_i at time t induces a cost equal to $C_{i,t}$. We also suppose that the time-windows $F(i), i = 1..n$, are stable with respect to the $<<$ precedence relation (see Section III.B). Then, the *Minimal Cost Homogeneous Non-Idling Scheduling* Problem comes as follows:

NON-IDLE$_c$: *Compute a feasible schedule T of $I = (J, F, <<, m)$ with a minimal cost $\sum_i T(i).C_{i,t}$*

5.1   A Linear Formulation of the NON-IDLE$_c$ Problem

The rewriting scheme of Section II may be used in order express the NON-IDLE$_0$ while using LIP *(Linear Integer Programming)* formulation as follows:

- The main vector is clearly a $0,1 - vector\ z$, with indexation on $\{1..n\}.N$, whose meaning is, for every pair (i,t) *such that* $t \in F(i) : z_{i,t} = 1$ if job J_i is performed at instant $t = 1..TS$;
- In order to use the reformulation scheme of Theorem 1, we also consider a load integral vector $y = (y_t, t \in N)0$, whose meaning is that exactly y_t jobs are performed at instant t.

We derive from Theorem 1 that Vectors z and y define a feasible solution of NON-IDLE$_0$ if:

- For any job $J_i, \sum_t z_{i,t} = 1$;   (E5)
- For any instant $t \in \mathbf{N}, \sum_x z_{i,t} = y_t m$;   (E6)
- For any pair of jobs J_i, J_j such that $J_i << J_j, \sum_t t.z_{i,t} \sum_t t.z_{j,t}$;   (E7)
- The function $t \to y_t$ has a *pyramidal shape. (Pyramidal Shape Constraint)*

In order to express in a linear way the Pyramidal Shape constraint, we introduce 2 auxiliary vectors $\{0,1\} - vector\ V = (V_t, t \in N)$ and $W = (W_t, t \in N)$, whose meaning is:

- If $V_t = 1$ then y is non decreasing in t, that means $y_{t+1}y_t$;
- If $W_t = 0$ then y is no increasing in t, that means $y_t y_{t+1}$.
  Then the Pyramidal Constraint becomes:
- For any $t = 0..TS - 1, V_t - (y_{t+1} - y_t)1$ and $W_t - (y_{t+1} - y_t)0$;   (E8)
- For any $t = 0..TS - 1, V_t + W_t = 1$;   (E9)
- For any $t = 0..TS - 1, V_t V_{t+1}$.   (E10)

The last one of the 3 constraints above (constraint (E10)) ensures that the function $t \to V_t$ is non increasing, which means that it starts with values equal to 1 (y is non decreasing in t) and next turns to 0 values (y becomes non increasing).
   Then the global linear program of the *Non Idling Problem with Minimal Cost* NON-IDLE$_c$ comes as follows:

NON-IDLE$_c$ Linear Program :
{   *Unknown* vectors:

- $z = (z_{i,t}, i \in 1..n, t \in \mathbf{N}$ such that $t \in F(i))$ with $\{0,1\}$ values ;
- $y = (y_t, t \in \mathbf{N})0$, Integral;
- $V = (V_t, t \in \mathbf{N})$ and $W = (V_t, t \in T)$ with $\{0,1\}$ values ;
  Constraints :
- For any job $J_i, \sum_t z_{i,t} = 1$;
- For any instant t in $\mathbf{N}, \sum_i z_{i,t} = y_t m$;
- For any pair of jobs J_i, J_j such that $J_i << J_j, \sum_t t.z_{i,t} \sum_t t.z_{j,t}$;
- For any $t = 0..TS - 1, V_t - (y_{t+1} - y_t)1$ and $W_t - (y_{t+1} - y_t)0$;
- For any $t = 0..TS - 1, V_t + W_t = 1$;

- For any $t = 0..TS - 1, V_t V_{t+1}$.
  Minimize: $\sum_{i,t} C_{i,t}.z_{i,t}$ }

The algorithms which we are now going to propose in order to deal with NON-IDLE$_c$ will use part of this LIP formulation. More specifically, we shall take advantage from the following property:

Theorem 6: *The vertices of the Polyhedron which is induced by the constraints (E5) and (E6) are integral.*

Sketch of the proof: We know that, if y is fixed, the polyhedron $\mathbf{P}(y)$ induced by the constraints (E5) and (E6) and which is related to the vector z has integral vertices (total unimodularity of bipartite graph edge/vertex incidence matrices). Moreover, the constraints on y which ensure the feasibility of the constraints (E5) and (E6) may be written, as a consequence of Duality: (E11)

- $\sum_t, y_t = n;$
- For any $t, y_t m;$
- For any interval Ω of $\mathbf{N}$, $\sum_{t \in \Omega} y_t Card(\{i = 1..n,$ such that $F(i) \subseteq \Omega\}).$

Any vertex of the polyhedron which is defined by (E11) is integral (total unimodularity of the *Interval or consecutive one* matrices). Then it only simple computation to check that if (z, y) is a feasible rational solution of (E5), (E6), (we relax the integrality constraints on z and y) then:

- y satisfies (E11), and so is a convex combination of vertices $v_1..v_k$ of the polyhedron defined by (E11);
- consequently, z may be decomposed into the same convex combination of rational solutions $z(v_1)..z(v_k)$ of $P(v_1)..P(v_k);$
- by transitivity, one gets that (z, y) may be decomposed as a convex combination of integral solutions of (P5), (P6) and this yields the result.

Before starting with the algorithms, we may notice that:

Remark 1: *the case when the cost coefficients $C_{i,t}, i = 1..n, t \in \mathbf{N}$. are monotonic.*

We say that the cost coefficients $C_{i,t}, i = 1..n, t \in \mathbf{N}$. are *monotonic* if, for every pair J_i, J_j of jobs such that $J_i << J_j$, and every time unit pair t, t such that $t < t$, we have: $C_{i,t} + C_{j,t} C_{i,t} + C_{j,t}$. (E12)

If the cost vector C is *monotonic*, then we see that we may relax the (E7) constraint from the above linear program without deteriorating the solution, since, in case the vector z involves any pair J_i, J_j such that:

- $J_i << J_j;$
- J_i is done after J_i according to $z;$

we only have to use the stability of the time-windows $F(i), i = 1..n$, with respect to the $<<$ precedence relation and to switch J_i and J_j inside the schedule in order to make the precedence constraint related to J_i and J_j satisfied, while this operation does not deteriorate the cost of the schedule (because of (E11)). We deduce from Theorem 6 that, in such a case, relaxing the *Pyramidal Shape* property turns our problem into a linear program which may be solved by application of the Simplex algorithm.

Remark 2: *Eliminating the V, W vectors.*
The pyramidal shape property means that there exists t_0 such that:

- y_t is non decreasing on $\{1..t_01\}$;
- y_t is non increasing on $\{t_0..TS - 1\}$. So we may define the parametrized restriction NON-IDLE$_c(t_0)$ of the NON-IDLE$_c$ Program by setting:

NON-IDLE$_c(t_0)$ Linear Program:
{   *Unknown* vectors:

- $z = (z_{i,t}, i \in 1..n, t \in \mathbf{N})$ such that $t \in F(i)$, with $\{0, 1\}$ values ;
- $y = (y_t, t \in \mathbf{N})0$, Integral;
  Constraints :
- For any job $J_i, \sum_t z_{i,t} = 1$;
- For any instant t in N, $\sum_i z_{i,t} = y_t m$;
- For any pair of jobs J_i, J_j such that $J_i << J_j, \sum_t t.z_{i,t} \sum_t t.z_{j,t}$;
- For any t in $\{1..t_01\}, y_t y_{t+1}$;   (E13)
- For any t in $\{t_0..TS - 1\}, y_t y_{t+1}$;   (E14)

    Minimize : $\sum_{i,t} C_{i,t}.z_{i,t}$}

Then solving NON-IDLE$_c$ means scanning $\mathbf{N}$, solving NON-IDLE$_c(t_0)$ for every value t_0 in $\mathbf{N}$, and picking up the best one, according to the following general scheme:

Algorithmic Scheme NON-IDLE$_c$:
Input: the NON-IDLE$_0$ instance $I = (J, F, <<, m)$ and the cost vector C;
Output: the vectors z and y;
Main Loop:
For $t_0 = 1..TS - 1$,
Solve NON-IDLE$_c(t_0)$ and
compute the related optimal solutions $z(t_0), y(t_0)$;   (E15)
$(z, y) \leftarrow$ best pair $(z(t_0), y(t_0)), t_0 = 1..TS1$;
Denote by *Top* the related t_0 value;

5.2   A Lagrangean Scheme for the NON-IDLE$_c$ Problem

We are going to deal with the NON-IDLE$_c$ Problem while using a heuristic approach and using both previous remarks. That means that (because of Remark

2) we are going to deal with NON-IDLE$_c$ through the generic NON-IDLE$_c$ algorithmic scheme described above, and focus on the Solve instruction (E13), that means on the resolution of NON-IDLE$_c(t_0)$. And, because of Remark 1, we are going to deal with the NON-IDLE$_c(t_0)$ problem through Lagrangean Relaxation of the (E13, E14 and E7) constraints.

Relaxing (E13), (E14) and (E7) means introducing the Lagrangean multipliers $\lambda = (\lambda_{ij}, i, j$ such that $J_i << J_j)0, \mu = (\mu_t, t = 1..t_01))0, \nu = (\nu_t, t = t_0..TS1))0$, and defining the related Lagrangean quantity:

$$L(t_0, z, y, \lambda, \mu, \nu) = \sum_{i,t} c_{i,t} \cdot z_{x,t} - \sum_{i,j \text{ such that } J_i << J_j} \lambda_{ij} \cdot (\sum_t t.z_{j,t} - \sum_t t.z_{i,t}) - \sum_{t=1..t_0-1} \mu_t \cdot (y_{t+1} - y_t) - \sum_{t=t_0..TS-1} \nu_t \cdot (y_t - y_{t+1}).$$

Then the Lagrangean relaxation RL-NON-IDLE$_c(t_0, \lambda, \mu, \nu)$ of NON-IDLE$_c(t_0)$ comes as follows:

RL-NON-IDLE$_c(t_0, \lambda, \mu, \nu)$ Linear Program :
{   *Unknown* vectors:

- $z = (z_{i,t}, i \in X, t = 1..TS$, such that $t \in F(i))$ with $\{0, 1\}$ values ;
- $y = (y_t, t \in \mathbf{N})0$, Integral;
  Constraints :
- For any job $J_i, \sum_t z_{i,t} = 1$;
- For any instant $t, \sum_i z_{i,t} = y_t m$;
  Minimize: $L(t_0, z, y, \lambda, \mu, \nu)$ }

Because Theorem 6, relaxing the Integrality constraint on z and y and applying the Simplex algorithm to the resulting linear program RL-NON-IDLE$_c(t_0, \lambda, \mu, \nu)^*$ yields integral solutions and an optimal value VAL(t_0, λ, μ, ν) which is the optimal value of RL-NON-IDLE$_c(t_0, \lambda, \mu, \nu)$. So, we may apply the standard Lagrangean Relaxation scheme and compute Lag-VAL(t_0) = $Sup_{\lambda, \mu, \nu 0} VAL(t_0, \lambda, \mu, \nu)$, together with the related Lagrangean multipliers λ^*, μ^*, ν^*. As a matter of fact, Duality Theory tells us that, in order to do it, we only need to solve the Linear relaxation (relaxation of the integrality constraints on z and y) RL-NON-IDLE$_c(t_0, \lambda, \mu, \nu)^*$ of the program NON-IDLE$_c(t_0)$, whose optimal value is exactly equal to Lag-VAL(t_0) and consider λ^*, μ^*, ν^* as being equal to the components of the related dual vectors which are respectively associated with (E13), (E14) and (E7). Then the general resolution scheme for NON-IDLE$_c(t_0)$ becomes:

Solve- NON-IDLE$_c(t_0)$ Procedure:
Solve (Simplex Algorithm) the Linear relaxation
RL-NON-IDLE$_c(t_0, \lambda, \mu, \nu)^*$ of NON-IDLE$_c(t_0)$);
Let λ^*, μ^*, ν^* be the components of the related dual vector which are
respectively associated with (E13), (E14) and (E7);
Let z and y be the 2 resulting integral vectors;
Turn (**Projection** scheme) z and y into a feasible solution of
NON-IDLE$_c(t_0)$;   (E16)

5.3 Instruction (E16): The Projection Scheme

It proceeds in 2 steps.

Projection Procedure:
Input: z and y obtained through resolution of the Linear relaxation RL-NON-IDLE$_c(t_0, \lambda, \mu, \nu)^*$ of NON-IDLE$_c(t_0)$, which may violate precedence constraints and the Pyramidal Shape constraint;
Output: z and y which are feasible in the sense of the NON-IDLE$_0$ problem;

1 th Step: we make the (E7) constraint become satisfied by switching pairs J_i, J_j such that:

- $J_i << J_j$;
- J_i is done after J_j according to z;

We do it as follows:
Not Stop;
While Not Stop do
Pick up J_i, minimal for the <<relation, such that there exists J_j such that:

- $J_i << J_j$;
- J_i is done at time t and J_i is done at time t according to z;
- $t < t$;

If J_i does not exist then Stop Else Schedule J_i, in t and J_i in t;
Clearly, in case the cost vector C is monotonic, this part of the process does not deteriorate the cost of the solution (z, y).

2 th Step: Then we make (E13) and (E14) satisfied by filling the $k - holes$ which may be induced in the schedule defined by the vector z: in order to do it, we simply adapt the feasibility algorithm SEARCH-SCHEDULE of Section IV, designed for the handling of NON-IDLE$_0$ and based on the use of propagation paths, while using the schedule defined by z as initial schedule.

5.4 Numerical Experiments

We performed tests while using the CPLEX library on a PC AMD opteron 2.1GHz, while using gcc 4.1 compiler. and used randomly generated instances whose size allowed direct computation of the optimal solution of the NON-IDLE$_c$ linear program. More specifically, the size product $n.\text{Card}(\mathbf{N})$ was included between 100 and 1000. As a matter of fact, in order to avoid too many instances which would not admit any feasible solution, we generated instances as follows :

- Generate a pyramidal shape vector y and a job set J with $n = \sum_t y_t$ jobs, together with a schedule vector z;
- Generate time-windows $F(i), i = 1..n$, and $<< -pairs(J_i << J_j)$, in such a way that z be a feasible schedule for the NON-IDLE$_0$ related instance; (E17)

- Make the time-windows $F(i), i = 1..n$, be stable (see Section III.B) with respect to the relation $<<$;
- Generate cost vector C;

We use the medium size α of the time-windows $F(i), i = 1..n$, and the number β of $<< -pairs$ as parameters of this process. We manage in such a way cost C may be monotonic. While performing those experiments, we focus on:

- the gap $GAP - LAG$ between the value $Lag - VAL(Top)$ and the optimal value $OPT - VAL$ of NON-IDLE$_c(Top)$, where Top is the optimal t_0 value of the NON-IDLE$_c$ algorithmic scheme;
- the numbers $NS-$ $<<$ and $HOLE$ of respectively (E7) and (E13, E14) violated constraints, and on the gap GAP between the optimal value of NON-IDLE$_c(Top)$, and the value of the solution (z, y) obtained at the end of the above process, that means after application of the *Projection* process;
- the impact of the *monotonicity* property.

Here are the results which were obtained on ten instances generated this way (Id denotes the instance):

Table 1. Performance Analysis of the NON-IDLE$_c(t_0)$ Lagrangean relaxation scheme

Id	Monoton	n.TS	m, α, β	GAP − LAG(%)	GAP(%)	HOLE, NS− <<
1	Yes	20.10	3, 4, 10	0	0	0, 0
2	No	20.10	3, 4, 10	1.8	4.1	0, 2
3	Yes	20.10	5, 5, 20	0	0	0, 0
4	No	20.10	5, 5, 20	5.2	10.5	1, 7
5	Yes	30.10	3, 4, 10	0	0	0, 0
6	No	30.10	3, 4, 10	2.8	7.1	1, 5
7	Yes	30.30	5, 5, 10	0.6	2.1	2, 0
8	No	30.30	5, 5, 10	1.8	3.7	0, 2
9	Yes	30.30	5, 8, 40	1.0	2.5	3, 0
10	No	30.30	5, 8, 40	3.9	9.6	3, 6

Comment: we notice that, in most cases, the difference between the value $Lag - VAL(t_0)$ and the optimal value of NON-IDLE$_c(t_0)$ is very small, and that, in many case, the y vector which derives from the Lagrangean relaxation process has a *Pyramidal Shape*. Still, in case there are many $<< -pairs$ and the cost vector C is non monotonic, the projection process should have to be improved.

6   Conclusion

In this paper, we have just been studying a variant of the *homogeneous m-machine non-idling* problem, where weakly dependent unit-jobs have to be scheduled within their time windows so that the non-idling constraints must be satisfied not only for each machine but for every subset of machines. A structural necessary and sufficient condition for an instance to be feasible has been provided and an algorithm has been developed for the existence problem. This algorithm may be extended to the case when a job dependent cost function is associated with the termination of each job and the makespan has to be minimized. However, several important questions about the complexity of more general problems with the same HNI constraints (non unit jobs, linear costs) are still open. Also, it would be quite interesting to get the complexity status of the non-homogeneous variant of the problem (more frequent when it comes to applications) which corresponds to the case when the non-idling constraint has only to be satisfied on each machine or on a given subset of machines.

References

[1] Baptiste, P.: Scheduling unit tasks to minimize the number of idle periods: a polynomial time algorithm for off-line dynamic power management; Research Report, Laboratoire d Informatique CNRS LIX (2005)
[2] Chretienne, P.: On single-machine scheduling without intermediate delays. Discrete Applied Maths 13-156, 2543–2550 (2008)
[3] Jouglet, A.: Single-machine scheduling with no-idle time and release dates to minimize a regular criterion. Journal of Scheduling 15(2), 217–238 (2012)
[4] Valente, J.M.S., Alves, R.A.F.S.: An exact approach to early/tardy scheduling with release dates. Computers and Operations Research 32, 2905–2917 (2005)
[5] Landis, K.: Group technology and cellular manufacturing in the Westvaco Los Angeles VH Department, Project Report in IOM 581, School of Business, University of Southern California (1983)
[6] Irani, S., Pruhs, K.: Algorithmic problems in power management, vol. 36, pp. 63–76. ACM Press, New York (2005)

Flow Models for Project Scheduling with Transfer Delays and Financial Constraints

Alain Quilliot and Hélène Toussaint

LIMOS, UMR CNRS 6158, Bat. ISIMA, Universit BLAISE PASCAL,
Campus des Czeaux, BP 125, 63173 Aubiere, France
{alain.quilliot,helene.toussaint}@isima.fr

Abstract. This paper deals with two extensions of the *Resource Constrained Project Scheduling Problem* (RCPSP), which involve resource transfer delays and "Financial" resources. Flow models are used in order to formalize those extended RCPSP, which contain the standard RCPSP and lead us to the *Timed Flow Polyhedron* and to several structural results. This framework gives rise to generic Insertion operators, as well as greedy/local search algorithms. We end with numerical tests.

Keywords: Resource Constrained Scheduling, Network Flow Theory.

1 Introduction

Dealing with *Resource Constrained Project Scheduling Problems* (RCPSP: see [1,2,3]) means scheduling tasks, submitted to temporal and resource constraints, while minimizing the induced *Makespan* value. This problem has been extensively studied: [4,5,6,7]; its theoretical analysis requires sophisticated mathematical tools: linear programming, posets, hypergraphs..: [8]. While Standard RCPSP only involves deterministic non pre-emptive tasks and renewable resources, extended models adress pre-emption: [9], time lags: [10], non renewable resources: [11,12], non constant profiles, robustness: [13], deadlines and penalties, redundant resources: [14,15]. A survey about RCPSP variants is available in [16].

RCPSP problems are usually NP-Complete, and getting exact results becomes hard as soon as there are more than 60 tasks and 4 resources: [17,18]. Exact methods are most often branch and bound, cut generation and constraint propagation based: [19,12,20]. Powerful lower bounds derive from column generation techniques applied to specific LP models, energetic reasoning processes or largest paths computing: [25, 26]. But efficient heuristics may be designed: greedy algorithms based on priority rules or insertion techniques: [13,23,24], local search methods: [25,26]. Dynamic RCPSP is most often handled through priority rule based algorithms: [24].

This paper studies two extensions of the RCPSP Problem. The first one involves *Resource Transfer Delays* and is denoted by RCPSTDP: resources are transmitted from one task to another, and those communication tasks involve

S. Fidanova (Ed.): *Recent Advances in Computational Optimization*, SCI 470, pp. 135–154.
DOI: 10.1007/978-3-319-00410-5_8   © Springer International Publishing Switzerland 2013

delays. Such a RCPSPTDP model corresponds to the case when every task takes place on a given production unit and when part of the resources (workers, equipments) need to be transported from one unit to another. Coping with such a problem requires the existence of an explicit representation of the way resources transit from one production unit to another and, thus, leads us to make appear a *Network Flow* component into our model. The second one involves a specific renewable resource, called *Financial Resource*, and is denoted by RCPSFRP. It is handled through the same framework, and through *Invest/Borrow* strategies (2 th RCPSFRP).

Network Flow Theory: [27], is devoted to problems which involve the circulation of goods, people, energy,.... It was used in order to model transportation, telecommunication and energy distribution systems: [27]. The existence of a link between the RCPSP and Network Flow Theory has already been noticed in [23,24,6,14], and been used in order to get ILP formulations and insertion algorithms. Still, few works have explicitly involved the network flow machinery into the design of generic algorithms. So, we shall first explain the way RCPSTDP and RCPSFRP may be cast into the Network Flow framework. Next, we shall state structural results about connectivity and cut management. Finally, we shall derive from this theoretical work generic *insertion* mechanisms, close to the insertion mechanisms which were proposed in [23,24], and use them in order to design and test greedy and local search algorithms.

2   Network Flow Model Related to a RCPSTDP Instance

Preliminary Notations and Definitions. We denote by $\leftarrow$ the value allocation operator: $x \leftarrow \alpha$ means that variable x takes value α. $\mathbb{Q}$ is the rational number set. If τ is some partial order relation, then $\tau^=$ is the relation (τ or $=$) and $Tr(\tau)$ is the *transitive closure* of τ. An oriented graph (network) G with node set Z and arc set E is denoted by $G = (Z, E)$. An arc e with origin/destination nodes x and y is denoted by (x, y). A partial graph (sub-graph) of G is the restriction of G to some subset of $E(Z)$.

2.1   The Non Preemptive RCPSTDP Problem

A Standard RCPSP instance $\mathcal{I} = (V, K, R, r, d, \varphi)$ is defined by:

- a set V of non pre-emptive tasks: $d_v > 0$ is the duration of task v;
- a binary no circuit *precedence* relation φ, defined on V: $v\varphi w$ means that v must be over before w starts;
- a finite renewable *resource* set K: R_k is the initial available amount of resource $k \in K$; task v requires $r_{k,v}$ resource k, and, once it is over, gives this resource back.

Solving $\mathcal{I}$ means computing, for any $v \in V$, its starting time $T_v \geq 0$, in such a way that:

- if v and $w \in V$ are such that $v\varphi w$, then $T_v + d_v \leq T_w$; (Precedence Constraint)
- at any time $t \geq 0$, and for any resource $k \in K$, $\sum_{v \in U(T,t)} r_{k,v} \leq R_k$, with $U(T,t) = \{v \in V$ such that $T_v \leq t < T_v + d_v\} \subseteq V$ is the set of the tasks which are running at time t; (Resource Constraint)
- the *makespan* $Makespan(T) = Sup_{v \in V}(T_v + d_v)$ is the smallest possible.

An instance $\mathcal{I}_{TD} = (V, K, R, r, d, \varphi, Lag, Depot)$ of the *Resource Constrained Project Scheduling with Transfer Delays Problem* (RCPSTDP) is defined as above, while taking into account the *Delay* function Lag: tasks of V are run at different places inside some *production* space, and resources circulate. At time 0, resources are all located at a same place $Depot$, and they must be back to $Depot$ for the project to be over. Then the *Delay* $\mathbb{Q}$-valued function Lag, associates, with any pair $v, w \in V \cup \{Depot\}$, a value $Lag(v, w) \geq 0$: if task v (or $Depot$) transmits some resource to task w (or to $Depot$) or if $v\varphi w$ (v transmits some output to w which uses it as an input), then transferring this resource requires a $Lag(v, w)$ delay between the ending time of v (0 if $v = Depot$) and the starting time of w (Delay Constraint). So, solving $\mathcal{I}_{TD}$ means simultaneously computing a Time vector T as in the simple case, and an ad hoc description F of the way resources are provided to the tasks.

Remark 1. RCPSTD and RCPSP with *Time Lags* are quite different problems, since delays $lag(v, w)$ only impact tasks v, w which exchange resources.

2.2 Linking Network Flows with RCPSTDP: Timed Flows

We formalize now the description of the way resources are provided to the tasks.

Recall: Network Flows. Given a network $G = (Z, E)$, i.e. an oriented graph with node (vertex) set Z and arc set E, together with a $\mathbb{Q}$-valued function ϕ defined on the node set Z; a $\mathbb{Q}$-valued E-indexed vector f is a $\mathbb{Q}$-*flow vector* iff:
$\forall z \in Z, \sum_{z \text{ is the origin of } e} f_e = \sum_{z \text{ is the destination of } e} f_e$.
   If I is some *commodity* set, if $\phi = (\phi(i), i \in I)$ is a *commodity function*, i.e., if every $\phi(i), i \in I$, is a $\mathbb{Q}$-valued function defined on the node set Z, then we call ϕ-*flow vector* any collection $f = (f(i), i \in I)$, where every $f(i), i \in I$, is a $\phi(i)$-*flow vector*.

The Activity Network. Let $\mathcal{I}_{TD} = (V, K, R, r, d, \varphi, Lag, Depot)$ be a RCP-SPTD instance. We derive from $\mathcal{I}_{TD}$ the *Activity Network* $\mathcal{N}(V) = (V^*, E^*)$ by introducing two auxiliary tasks $Start$ and End, and by setting:

- $V^* = V \cup \{Start, End\}$ = node set of $\mathcal{N}(V)$;
- $E^* = \{(v, v'), v, v' \in V\} \cup \{(Start, v), v \in V \cup \{End\}\} \cup \{(v, End), v \in V \cup \{Start\}\}$ = arc set of $\mathcal{N}(V)$.

We define the E^*-indexed *length vector* d^* by setting:

- for any $v \in V, d^*_{(Start,v)} = Lag(Depot, v)$ and $d^*_{(v,End)} = d_v + Lag(v, Depot)$;
- $d^*_{(End,Start)} = -\infty$;
- for any $v \in V, w \in V \cup \{End\}, d^*_{(v,w)} = d_v + Lag(v, w)$.

We provide the node set V^* with a *commodity vector* r^*, by setting, for every resource $k \in K$ and for any $v \in V^*$: if $v \in V$ then $r^*_{k,v} = r_{k,v}$ else $r^*_{k,v} = R_k$. We define the *precedence arc subset* E^*_φ by setting: $E^*_\varphi = \{(v, v'), v, v' \in V$ such that $v \, Tr(\varphi) \, v'\} \cup \{(Start, v), (v, End), v \in V\}$, where $Tr(\varphi)$ is the *transitive closure* of the φ relation.

Feasible Solutions of $\mathcal{I}_{TD}$ and r^*-Flow Vectors. Then, a representation of the circulation of resources between the tasks of V, consists into a r^*-flow vector $F = (F(k), k \in K)$, defined on the Activity Network $\mathcal{N}(V)$, which may be viewed as transporting the resources $k \in K$, from $Depot$ (the source-node $Start$) to $Depot$ (the end-node End), while providing the tasks $v \in V$ with the required resources. The support arc subset $E(F, \varphi)$ of F is: $E(F, \varphi) = E^*_\varphi \cup \{(v, w), v, w \in V$ such that $F_{(v,w)} \neq 0\}$. Clearly, $E(F, \varphi)$ has *no circuit*, (we say F is *no circuit*). We easily see (see for instance [23,24]), that even if we only deal with a simple RCPSP instance $\mathcal{I} = (V, K, R, r, d, \varphi, Lag)$, we may derive, as described in Figure 1, such a flow F from any feasible schedule T.

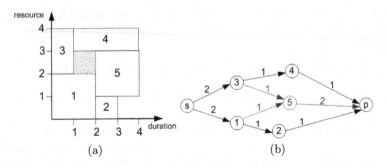

Fig. 1. (a) Gantt chart - (b) Flow representation

Casting RCPSPTDP into a Formal Framework: *Timed Flows*. Let F be a no circuit r^*-flow vector and T a time V-indexed vector as above. If we extend the time vector T to $V \cup \{Start, End\}$ by setting: $T_{Start} = 0$; $T_{End} = \Delta = Makespan(T) = Sup_{v \in V}(T_v + d_v + Lag(v, Depot))$, then, for any arc $e = (v, w)$ in E^*, we get the implication:

$$e = (v, w) \in E(F, \varphi) \Rightarrow (T_w \geq T_v + d_v \Leftrightarrow T_w \geq T_v + d^*_e) \quad \text{(P1)}$$

This leads us to define a *Timed (r^*, d^*)-Flow* as being any such pair (F, T) made of a no circuit r^*-flow vector F and a time vector T such that (P1) is true. One easily checks that any such a *Timed (r^*, d^*)-Flow(F, T)* defines a feasible solution of $\mathcal{I}_{TD}$. It allows us to reformulate RCPSTDP as follows:

RCPSPTD Timed Flow Reformulation. *Solving the RCPSTDP instance* $\mathcal{I}_{TD} = (V, K, R, r, d, \varphi, Lag, Depot)$ *means computing, on the Activity network* $\mathcal{N}(V)$, *a Timed* (r^*, d^*)*-Flow* (F, T) *such that* T_{End} *is the smallest possible.*

Also, following ([23,24]) one easily checks that:

Theorem 1: Standard RCPSP Reformulation Theorem. *Any feasible solution* T *of the Standard RCPSP instance* $\mathcal{I} = (V, K, R, r, d, \varphi)$ *may be extended into a feasible solution* (F, T) *of the RCPSTDP instance* $\mathcal{I}_{TD} = (V, K, R, r, d, \varphi, 0, Depot)$.

2.3   Connectivity Theorem

The efficiency of the flow machinery derives from properties of the flow polyhedron. Thus, one may ask about the polyhedron of the no circuit r^*-flow vectors.

The No Circuit r^*-Flow Polyhedral Vertex Set. r^*-flow vectors $F \geq 0$ defined on the network $\mathcal{N}(V)$ defines a bounded polyhedron $\mathcal{P}_{r^*}$. F is a vertex of $\mathcal{P}_{r^*}$ which contains *no non null alternated cycle*, that means no cycle $(v_0, v_1, \ldots, v_n = v_0)$ such that:

- n is even and all the nodes $v_0, \ldots, v_{n-1}$ are distinct (the cycle is elementary);
- there exists $k \in K$ such that:
  - the arcs $(v_0, v_1), (v_2, v_3), \ldots, (v_{n-2}, v_{n-1})$ are all endowed with non null $F(k)$ values;
  - the arcs $(v_2, v_1), (v_4, v_3), \ldots, (v_0, v_{n-1})$ are all endowed with non null $F(k)$ values.

We denote by $\mathcal{S}_{r^*}$ the vertex set of this polyhedron. This vertex set is endowed with a canonical *adjacency* relation $\mathcal{R}$, which may be characterized as follows:

- let Γ be some even cycle $(v_0, v_1, \ldots, v_n = v_0)$ in $\mathcal{N}(V)$: the *alternated cycle flow* f^Γ is defined by:
  - $f_e^\Gamma = +1$ for any arc $e = (v_0, v_1), (v_2, v_3), \ldots, (v_{n-2}, v_{n-1})$;
  - $f_e^\Gamma = -1$ for any arc $e = (v_2, v_1), (v_4, v_3), \ldots, (v_0, v_{n-1})$.
- F, F' in $\mathcal{S}_{r^*}$ are $\mathcal{R}$-adjacent if there exists some resource $k_0 \in K$, some even cycle Γ and some number $\lambda \geq 0$, such that we have: for any $k \neq k_0, F(k) - F'(k) = 0$; $F'(k_0) - F(k_0) = \lambda.f^\Gamma$. In such a case, value λ is unique, and F' derives from F through *redirection* of $F(k)$ on Γ.

It comes from LP Theory that $\mathcal{S}_{r^*}$ is connected for the relation $\mathcal{R}$. Since we deal here with no circuit r^*-flows, we may ask whether the set $\mathcal{SN}_{r^*}$ of vertices of $\mathcal{P}_{r^*}$ which define no circuit r^*-flow vectors have this connectivity property. We call $\mathcal{SN}_{r^*}$ *the No Circuit r^*-Flow Polyhedral Vertex Set*. Then we state:

Theorem 2: Connectivity Theorem. *If we suppose that, for any* $k \in K, v, w \in V$, *we have:* $r_{k,v} + r_{k,w} \leq R_k$ *(Parallelism Hypothesis), then the No Circuit r^*-Flow Polyhedral Vertex Set* $\mathcal{SN}_{r^*}$ *is connected for the canonical adjacency relation* $\mathcal{R}$.

Comment. thus, one may handle RCPSP instance through flow local search.

Proof of theorem 2 (Sketch of the Proof). We define a linear r^*-flow vector as a no circuit r^*-flow vector $F \geq 0$ such that the transitive extension of the support set $E(F, \varphi)$ is linear. We denote by $\mathcal{SNL}_{r^*}$ the subset of $\mathcal{SN}_{r^*}$ made with linear r^*-flow vectors. If σ is a linear ordering of $V \cup \{Start, End\}$, compatible with φ, we denote by $\mathcal{SN}_{r^*}(\sigma)$ the subset of $\mathcal{SN}_{r^*}$ which corresponds to the case when σ is as a linear extension of the transitive extension of $E(F, \varphi)$. Then we check, by using ad hoc flow redirection processes, that:

Lemma. $\mathcal{SN}_{r^*}(\sigma)$ is connected for the $\mathcal{R}$ relation. Also, if, for any $v \in V, k \in K, r_{k,v} \neq 0$ and if the parallelism holds, then we may state that:

- If F and F' are in $\mathcal{SNL}_{r^*}$, then there exists a $\mathcal{R}$-path from F to F'; (P2)
- for any linear ordering σ of $V \cup \{Start, End\}$, which is compatible with φ, the intersection of $\mathcal{SNL}_{r^*}$ and $\mathcal{SN}_{r^*}(\sigma)$ is non empty.   (P3).

This lemma allows us to conclude to the $\mathcal{R}$-connectivity of $\mathcal{SN}_{r^*}$ when, for every $v \in V$ and every $k \in K$, the quantity $r_{k,v}$ is non null. In order to get our result in the general case, we use a trick which involves topology. Let $\delta > 0$ be a small positive number. For every activity v and any resource k, such that $r_{k,v} = 0$, we replace $r_{k,v}$ by δ, and R_k by $R_k + Card(V(k)).\delta$, where $V(k) = \{v \in V$ such that $r_{k,v} = 0\}$. We denote by $\mathcal{SN}_{r^*}^{\delta}$ the respective related polyhedron vertex sets and by R^{δ} the related adjacency relation. It comes from above that $\mathcal{SN}_{r^*}^{\delta}$ is connected for the relation $\mathcal{R}^{\delta}$. Also, we see that if F is some vertex in $\mathcal{SN}_{r^*}$, then the r^*-flow vector F^{δ} defined by:

- for any v and any k such that $r_{k,v} = 0, F^{\delta}(k)_{(Start,v)} = \delta = F^{\delta}(k)_{(v,End)}$;
- $F^{\delta}(k)_{(Start,s)} = Card(V(k)).\delta$;
- for any other pair $(e,k), k \in K$, e in the arc set of the network $\mathcal{N}(V)$, $F^{\delta}(k)_e = F(k)_e$;

is no circuit, does not admit any non null alternated cycle, and so is in $\mathcal{SN}_{r^*}^{\delta}$. We conclude by checking that any pair F, H of elements of $\mathcal{SN}_{r^*}$, may be connected by a path Γ which is the limit, when δ converges to 0, of some path sequence $\Gamma^{\delta}, \delta > 0$, where every path Γ^{δ} connects F^{δ} and H^{δ} in $\mathcal{SN}_{r^*}^{\delta}$. End-Proof.

Remark. one easily check that the Parallelism hypothesis cannot be removed.

3   Insertion Scheme, Insertion Problem and Algorithms

As told in 2, *Timed Flow* formalism aims at the application of ad hoc network flow algorithmic tools to RCPSTDP instances. So section 3 describes the way it can be done. Basically, our RCPSTDP algorithms perform *insertion/removal* processes which may be compared with those which have been proposed in [23,24] for standard RCPSP: the basic difference lays upon the fact that every time the insertion/removal of some activity is performed, it involves the resolution of a specific *Insertion-Flow* sub-problem related to a given *Cut* of the currently inserted task set: the related resolution process updates all the flow values which express the flow transportation between both sides of this *Cut*, and implement the *Connectivity Theorem* of Section 2. More precisely, at any time during the process of a RCPSTDP instance $\mathcal{I}_{TD} = (V, K, R, r, d, \varphi, Lag, Depot)$, we are provided with an *Inserted Activity subset* W of V, with a no circuit r^*- flow vector F defined on the *Activity Network* $\mathcal{N}(W)$, and with two $\mathbb{Q}$-valued time vectors T and $T^* \geq 0$, both with indexation on W^*, in such a way that, for any $v \in W^*$:   (P4)

- T_v = Length of a largest path from *Start* to v in the *Support Partial Activity Network* defined by $E(F, \varphi)$, for the length vector d^*;
- T_v^* = Length of a largest path from v to *End* in the *Support Partial Activity Network* defined by $E(F, \varphi)$, for the length vector d^*.

Clearly, (F, T) is a timed (r^*, d^*)-flow on $\mathcal{N}(W)$. Performing an *Insertion* means picking up some task v_0 in $V - W$, computing some *Cut*, i.e. a partition of W into 2 subsets U and $W - U$, such that no flow goes from to $(W - U) \cup \{End\}$ to $U \cup \{Start\}$, and turning (F, T), through the resolution of the *Insertion-Flow* Problem, into a timed (r^*, d^*)-flow defined on $\mathcal{N}(W \cup \{v_0\})$, in such a way that v_0 receive flow values from $U \cup \{Start\}$ and give them back to $(W - U) \cup \{End\}$. Performing a *Removal* means reversing this operation. In order to better explain those mechanisms, we shall introduce the *Flow-Insertion Problem*. Meanwhile, we illustrate this mechanism through Fig. 2, which represents, (a), a partial solution with 4 tasks and a *Cut* ($U = \{1, 3\}, W - U = \{2, 4\}$), and, (b), the insertion of task $v_0 = 5$ (with $d_5 = 2$ and $r_5 = 2$) into this *Cut*.

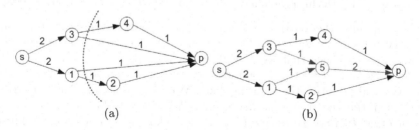

(a)   (b)

Fig. 2. (a) The flow and the *cut* (dotted line) - (b) the resulting flow after insertion

3.1   The Insertion Flow Problem

We say that an oriented graph $\mathcal{N} = (X, E)$ is almost-bipartite, if there exists some node z_0 in X such that the restriction of $\mathcal{N}$ to $X - \{z_0\}$ is bipartite, which means that $X - \{z_0\}$ may be written as the disjoint union $X - \{z_0\} = A \cup B$, of two disjoint independent sets A and B; let us suppose now that we are endowed with two positive (or null) $\mathbb{Q}$-valued A-indexed vectors Π, Out, with two positive (or null) $\mathbb{Q}$-valued B-indexed vectors Π^*, In, with some $\mathbb{Q}$-valued positive vectors Δ with indexation on $(A \cup \{z_0\}).(B \cup \{z_0\})$, and with a positive (or null) coefficient ρ such that: $\sum_{x \in A} Out(x) = \sum_{y \in B} In(y) \geq \rho$; then we say that a vector $G = (G_{x,y} \geq 0, x \in A \cup \{z_0\}, B \cup \{z_0\}) \geq 0$, is an *Insertion-Flow vector* related to $((X, E), z_0, A, B, In, Out, \Pi, \Pi^*, \Delta, \rho)$ iff:

- for any $x \in A$, $Out_x = \sum_{y \in B \cup \{x_0\}} G_{x,y}$;
- for any $y \in B$, $In_y = \sum_{x \in A \cup \{x_0\}} G_{x,y}$;
- $\rho = \sum_{y \in B} G_{z_0,y} = \sum_{x \in A} G_{x,z_0}$.

For such an *Insertion-Flow* vector G, we set:

- $Make1(G) = \text{Sup}_{x \in A, y \in B \text{ such that } (x,y) \in E \text{ or } G_{x,y} \neq 0}(\Pi_x + \Pi_y^* + \Delta_{x,y})$;
- $Make2(G) = \text{Sup}_{x \in A, y \in B \text{ such that } ((x,z_0) \in E \text{ or } G_{x,z_0} \neq 0 \text{ and } (z_0,y) \in E \text{ or } G_{z_0,y} \neq 0)}$
  $(\Pi_x + \Pi_y^* + \Delta_{x,z_0} + \Delta_{z_0,y})$;
- $I\text{-}Makespan(G) = \text{Sup}(Make1(G), Make2(G))$.

This definition leads us to introduce the following *Insertion-Flow Problem*: Given $(X, E), z_0, A, B, In, Out, \Pi, \Pi^*, \Delta, \rho$ as above. *Find a related insertion flow vector G in such a way that $I\text{-}Makespan(G)$ be the smallest possible.*

Explanation. if we refer to the previously described insertion process, and if $Card(K) = 1$, then we clearly see that we should think: (P5)

- $A = U \cup \{Start\}$; $B = (W - U) \cup \{End\}$;
- $E = (v, w)$, $v \in A, w \in B$, such that $v \, Tr(\varphi) \, w$;
- $z_0 = v_0$; $\rho = r_{k,v_0}$;
- for any v in $U \cup \{Start\}$, $\Pi_v = T_v$ and $Out_v = \sum_{w \in B} F(k)_{(v,w)}$;
- for any w in $B = (W - U) \cup \{End\}$, $\Pi_w^* = T_w^*$ and $In_w = \sum_{v \in A} F(k)_{(v,w)}$;
- for any pair v, w in $(A \cup z_0).(B \cup z_0)$, $\Delta_{v,w} = d_{(v,w)}^*$.

In order to deal with this problem, we design an *Insertion-Flow* procedure which:

- first computes the set Λ_{E,z_0} of the possible *attachment values*: the *attachment value* $u(G)$ of an *Insertion-Flow* vector G is given by: $u(G) = \text{Sup}(\Pi_x + \Delta_{x,z_0})$, $x \in A$ such that $(G_{x,z_0} \neq 0$ or $(x, z_0) \in E)$;
- next, **For** any $u \in \Lambda_{E,z_0}$, computes some *Insertion-Flow* G such that $u = u(G)$, and performs the following loop: **While** *Possible* **do**: Make decrease, through the search of some *Redirection Path*, the G value on the arcs (x, y) such that $I\text{-}Makespan(G) = G_{x,y}$, while making $u(G)$ remain unmodified;
- ends while keeping with the best *Insertion-Flow* G which was computed.

Theorem 3: Insertion-Flow Theorem. *The Insertion-Flow Procedure solves in an exact way the Insertion-flow Problem in polynomial time.*

Proof of theorem 3 (Sketch of the Proof). The key point is that $u(G)$ remains unmodified during the *While Possible* loop: then, a standard flow reasoning makes appear that if G is some *Insertion-Flow* vector, it $G(u)$ derives from the *While Possible* loop for a given u, and if we have both $u(G) = u$ and $I\text{-}Makespan(G(u)) > I\text{-}Makespan(G)$, then the cycle decomposition of the flow vector $G - G(u)$ makes appear some *Redirection Path* Γ which makes possible improving $G(u)$. *End-Proof.*

Fast-Try-Insertion and Lex-Insertion-Flow Procedures. Before insertion is effectively performed, it must be tried several times. We speed this process through a greedy approximation *Fast-Try-Insertion* heuristics, which set flow values on the arcs (x, y) which are the most likely to make increase *I-Makespan* in case $G_{x,y} = 0$. Also, we minimize the number of arcs (v, w) such that $(v \ (Not \ Tr(\varphi)) \ w$ and $F_{(v,w)} \neq 0)$ by turning *Insertion-Flow* into a *Lex-Insertion-Flow* procedure which tends to allocate G values to arcs of E, while E increases every time a new resource k is handled.

3.2   Generic Flow Algorithms for the RCPSTDP

A Greedy Insertion Algorithm RCPSTDP-Greedy-Flow. This algorithm works through successive insertions as explained in section 3. Clearly, a key point is about the computation of the *Cut U*. But searching for a best Cut U in the general sense seems to be a difficult problem:

Theorem 4: Cut Theorem. *The Best Cut problem as defined above is NP-Complete.*

Proof of theorem 4. Let us consider the following *Best Cut* Problem instance:

- there is only one resource we set $r_{k,x} = r_x$ and $R_k = R$;
- $\sum_{v \in W} r_v = R$; $r_{v_0} = R/2$; φ is the empty relation;
- for any $v \in W$, $d_v = 1$; $d_{v_0} = 2$.

Determining whether the optimal value of this *Best Cut* instance is no more than 2 means solving some *2-Partition* problem instance. *End-Proof.*

Still, if v is some node in $W \cup \{End\}$, we may set $Cut(v) = \{w \in W, w \neq v,$ such that $T_w \leq T_v\}$. So we can easily scan the set W, and choose $U = Cut(v)$ in such a way that an application of the *Insertion-Flow* Procedure yields the best possible *Makespan* value. The whole process **RCPSTDP-Greedy-Flow** , which may be randomized, comes as an application to the empty task set of the following **Packet-Insertion** Procedure:

A Local Search RCPSTDP-LS-Flow Algorithm. The above *Packet-Insertion* operator gives rise in a generic way to a local search operator *Transform-Insertion*. The idea is that, once we are endowed with a timed (r^*, d^*)-flow (F, T), we may pick up some (small) subset S of V, (take it away from V applying some *Reverse-Insertion* procedure) and, next, come back to inserting the activities of S into the pair (F, T). *Transform-Insertion* operates on any timed (r^*, d^*)-flow (F, T), through a parameter $S \subseteq V$:

Transform-Insertion(S): Reverse-Insertion(S); Packet-Insertion($V - S$).

Provided with this operator, we design local search algorithms, while picking up S as the task subset defined by a critical path (*Crit-Path* strategy) or by tasks which are simultaneously running at some critical time t (*Antichain* strategy).

Algorithm 1: Packet-Insertion Procedure

Input: the RCPSTDP instance $\mathcal{I}_{TD} = (V, K, R, r, d, \varphi, Lag, Depot)$, a subset W of V, a related no circuit r^*-flow vector F, and two related vectors T and T^* as in (P4);

Output: a timed (r^*, d^*)-flow (F, T), and the related *Makespan* value Δ;

Initialization: $S \leftarrow V - W$; $S_{Aux} \leftarrow S$;

while $S_{Aux} \neq Nil$ **do**

    Randomly Pick up v_0 in S_{Aux} and Remove it from S_{Aux} ;

    Compute v_1 in $W \cup \{End\}$, such that the application of the *Fast-Try-Insertion-Flow* procedure to:

- $X = W \cup \{Start, End, v_0\}$; $A = Cut(v_1) \cup \{Start\}$; $B = X - A - \{v_0\}$;
- $z_0, In, Out, \Pi, \Pi^*, \Delta, \rho, E = \{(x, y), x \in A \cup \{v_0\}, y \in B \cup \{v_0\}\}$, as in (P5);

    yields the best possible *Makespan* value;

    Let u_1 be the related attachment value: **Apply** *Lex-Insertion-Flow* to $Cut(v_1)$ and u_1, and perform the insertion of v_0 into the timed (r^*, d^*)-flow (F, T) in an effective way (update F, T and T^* values);

    $W \leftarrow W \cup \{v0\}$;

---

4   Numerical Tests on RCPSTDP

We performed experiments, on PC AMD opteron 2.1GHz, gcc 4.1 compiler. We tried instances from the PSPLIB testbed. Since we were not provided with optimal values for general RCPSTDP, we first tested the case $Lag = 0$, when RCPSP and RCPSPTD are the same, as well as ad hoc instances, with Lag such that the optimal values of both problems were the same. For every instance, *N-Ac*, *N-Res*, *N-Re* respectively denote the numbers of tasks, resources and replications, and we computed:

- *Time* = CPU time in seconds for the *N-rep* replications;
- *Gap-LB* (%) = gap between our values and: in case of 30 job instances, the optimal value; in case of 60 and 120 job instances, the best lower bound.
- *Gap-TB* (%) = gap between our values and: in case of 30 job instances, the optimal value; in case of 60/120 job instances, the largest path lower bound.

4.1   Experiments on PSPLIB Instances with $Lag = 0$

Our models and algorithms may be used in order to deal with standard RCPSP instances, and it is interesting to test their efficiency in such a specific context. The following tables 1 and 2 provide us with average results for the algorithms **RCPSTDP-Greedy-Flow** and **RCPSP-LS-Flow**, related to the PSPLIB packages: 30 jobs, 60 jobs, 120 jobs, when $Lag = 0$. The induced results are very satisfactory, taken into account the genericity of our algorithms.

Table 1. *RCPSP-Greedy-Flow* procedure, Mean Results

N-Ac	N-res	N-re	Time (s)	Gap-TB	Gap-LB
30	4	100	0.63	1.87	1.87
30	4	1000	6.3	0.92	0.92
60	4	100	4.54	16.91	7.10
60	4	1000	53.04	15.37	5.79
120	4	100	29.6	52.33	21.32
120	4	1000	515	48.84	18.5

4.2 Instances such That RCPSP and RCPSTDP Optimal Values Are the Same

These tests involve *difficult* instances, which comes as follows: we start from a standard RCPSP instance $\mathcal{I}$ of PSPLIB and from almost optimal solution T. Then we randomly generate *Lag* values such that T remains a feasible RCPSTDP schedule. For every pair (v, w) of actions of $\mathcal{I}$, which are parallel according to T, we compute a maximal *Lag* value $Max\text{-}Lag(T, v, w)$ which is compatible with T, and generate *Lag* values with a mean ratio $Lag/Max\text{-}Lag$ which vary from 10% to 50% (difficult instances). Table 3, provides us with results related to such 30 job PSPLIB instances, distributed into 5 groups according to the value of the mean ratio $Lag/Max\text{-}Lag$. The replications number is 100.

Table 2. *RCPSP-Greedy-Flow*, 100 Replications, Mean Results, Ad Hoc PSPLIB Instances

Group-Instance	Gap (%)	Time (s)
1 - 10%	2,5	2,71
2 20%	3,21	2,79
4 40%	8,5	2,94
5 50%	14,5	3,12

Comment. Clearly, the largest is the $Lag/Max\text{-}Lag$ mean ratio, the most difficult are the instances. Still, results remain satisfactory.

5 RCPSP with Financial Resources

In this section we study the way flow models may be used in order to deal with non renewable resources. As a matter of fact, we consider a specific non renewable resource, called *Financial* resource, which may vary in a very significant way, since it is produced by some tasks and consumed by others.

An instance $\mathcal{I} = (V, K, R, r, d, \varphi, \Phi, \phi, \psi, \delta)$ of the *Resource Constrained Project Scheduling with Financial Resource Problem* (RCPSFRP) is going to be defined by the same components (V, K, R, r, d, φ) as in the standard case, augmented with an initial *Cash* amount Φ, with two *Financial Resource* function ϕ and ψ and with a delay function δ, whose meaning comes as follows:

- launching some task v of V requires some amount $\phi(v)$ of *cash* (*Financial Resource*), and achieving this task allows the project manager getting back some *Cash* amount $\psi(v)$, which becomes available $\delta(v)$ times unit after v has been delivered. In case $\psi(v)$ is null, the delay $\delta(v)$ is also null;
- at time 0, the initial *Cash* amount which is available is Φ;
- the project is considered as over when all the tasks have been performed and when all the cash due has been collected.

Then solving the RCPSFRP instance $\mathcal{I} = (V, K, R, r, d, \varphi, \Phi, \phi, \psi, \delta)$ means computing, for any task $v \in V$, its starting time $T_v \geq 0$ in such a way that:

- if v and $w \in V$ are such that $v\varphi w$, then $T_v + d_v \leq T_w$; (Precedence Constraint)
- at any instant $t \geq 0$, and for any resource $k \in K$, we have $\sum_{v \in U(T,t)} r_{k,v} \leq R_k$, where the subset $U(T,t) \subseteq V$ is defined as the set of the tasks which are under execution at time t: $U(T,t) = \{v \in V \text{ such that } T_v \leq t < T_v + d_v\}$; (Resource Constraint)
- at any instant $t \geq 0$, the available cash amount $\mathcal{M}(T,t)$ is non negative: $\mathcal{M}(T,t)$ may be computed through the following formula: $\mathcal{M}(T,t) = \Phi - \sum_{v \in UF(T,t)} \phi(v) + \sum_{v \in UF^*(T,t)} (\psi(v) - \phi(v))$, where the sets $UF^*(T,t)$ and $UF(T,t)$ are defined by: (Financial Constraint)
  - $UF(T,t) = \{v \in V \text{ such that } T_v \leq t < T_v + d_v + \delta(v)\}$;
  - $UF^*(T,t) = \{v \in V \text{ such that } T_v + d_v + \delta(v) \leq t\}$;
  The second component in the above formula represents the cash flow which has been generated by the tasks which have been achieved at time t, while the first one represents the cash flow which is immobilized at time t;
- the *Makespan* value $F\text{-}Makespan(T) = \text{Sup}_{v \in V}(T_v + d_v + \delta(v))$ is minimal.

Remark. Checking whether the feasibility of a RCPSFRP instance $\mathcal{I} = (V, K, R, r, d, \varphi, \Phi, \phi, \psi, \delta)$ is NP-Complete. One may check that, in case $\Phi = 1$ and V may be written as a bipartition $V = A \cup B$, in such a way that $\text{Card}(A) = \text{Card}(B)$ and:

- for any v in A, $\phi(v) = 1$ and $\psi(v) = 0$;
- for any v in B, $\phi(v) = 0$ and $\psi(v) = 1$;

then checking the feasibility of the related instance $\mathcal{I}$ contains the *Bandwidth Problem on Co-Bipartite Balanced Graphs*, which is NP-Complete (see [6]).

5.1   Adapting the Flow Model to a RCPSFRP Instance

We keep on with the same *Task Network* $\mathcal{N}(V) = (V^*, E^*)$ as in Section 2, and provide the arc set E^* with an additional *length vector* δ^* by setting:

- for any $v \in V$, $\delta^*_{(Start,v)} = 0$; $\delta^*_{(End,Start)} = 0$;
- for any $v \in V$, $w \in V \cup \{End\}$, $\delta^*_{(v,w)} = d_v + \delta(v)$.

We also define, on the node set V^*, two *financial commodity vectors* ϕ^* and ψ^*, by setting, for every $v \in V^* = V \cup \{Start, End\}$:

- if $v \in V$ then $\phi_v^* = \phi(v)$ and $\psi_v^* = \psi(v)$;
- $\psi_{Start}^* = \Phi$; $\phi_{Start}^* = 0$; $\psi_{End}^* = 0$; $\phi_{End}^* = \Phi - \sum_{v \in V} \phi(v) + \sum_{v \in V} \psi(v)$.

As in Section 2, we easily check that, with any feasible solution T of the RCPS-FRP instance $\mathcal{I} = (V, K, R, r, d, \varphi, \Phi, \phi, \psi, \delta)$, we may associate a pair $(F, F\text{-}cash)$, where:

- $F = (F(k)_{(v,w)}, k \in K, (v,w) \in E^*)$ is a r^*-flow vector;
- $F\text{-}cash = (F\text{-}cash_{(v,w)}, (v,w) \in E^*)$ is such that:
  - for any task $v \in V$, $\sum_{v \text{ is the origin of } e} F\text{-}cash_e = \psi(v)$;
    $\sum_{v \text{ is the extremity of } e} F\text{-}cash_e = \phi(v)$;
  - $\sum_{Start \text{ is the origin of } e} F\text{-}cash_e = \Phi = \psi_{Start}^*$;
  - $\sum_{End \text{ is the extremity of } e} F\text{-}cash_e = \phi_{End}^* = \Phi - \sum_{v \in V} \phi(v) + \sum_{v \in V} \psi(v)$;

  We say that $F\text{-}cash$ is a (ϕ^*, ψ^*)-flow vector.

Interpretation. the vector $(F, F\text{-}cash)$ transports the resources $k \in K$ and the *Cash* resource from *Start* to *End*, and provides the tasks $v \in V$ with those resources in such a way it allows them to be run.

Example. We consider 4 tasks A, B, C, D, with durations 2, 5, 3, 4, together with 1 resource. Resource requirements are given by the following table:

	A	B	C	D
R	3	1	2	4
Cash	(5,2)	(5,3)	(2,8)	(1, 12)

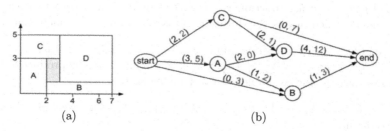

Fig. 3. A RCPSFRP Flow Solution: (a) Gantt chart - (b) Flow representation: the first number denotes the resource flow and the second one the Cash flow. Initial Cash is 10, initial resource amount is 5, payment delays are null

Then we set: $E2(F\text{-}cash) = \{(v, w), v, w \in V$ such that the flow value $F - cash_{(v,v')}$ is non null$\}$. This allows us to define the *support arc Subset* $E^*(F, F\text{-}cash, \varphi)$ of $(F, F\text{-}cash)$ by setting $E^*(F, F\text{-}cash, \varphi) = E(F, \varphi) \cup E2(F\text{-}cash)$, where $E(F, \varphi)$ is defined as in Section 2. Clearly, $E^*(F, F\text{-}cash, \varphi)$ is no circuit, and, if we extend T to $V \cup \{Start, End\}$ in a natural way by setting $T_{Start} = 0$ and $T_{End} = F\text{-}Makespan(T)$, we get, for any arc $e = (v, w)$ in the arc set E^*, the following implications:

- $(v, w) \in E(F, \varphi) \Rightarrow T_w \geq T_v + d_v \Leftrightarrow T_w \geq T_v + d_e^*;$ (P6)
- $(v, w) \in E2(F\text{-}cash) \Rightarrow T_w \geq T_v + d_v + \delta(v) \Leftrightarrow T_w \geq T_v + \delta_e^*.$ (P7)

We say that such a triple $(F, F\text{-}cash, T)$ which satisfies (P6) and (P7) defines a Timed (r^*, ϕ^*, ψ^*)-Flow, and we easily check that:

Theorem 5: Reformulation Theorem for the Financial Case. *Solving the RCPSFRP instance* $\mathcal{I} = (V, K, R, r, d, \varphi, \Phi, \phi, \psi, \delta)$ *means computing a Timed* (r^*, ϕ^*, ψ^*)-Flow $(F, F\text{-}cash, T)$ *such that* $F\text{-}Makespan(T) = Sup_{v \in V}(T_v + d_v + \delta(v))$ *is the smallest possible.*

Clearly, in case Φ, ϕ, ψ, δ are null, we find again the standard RCPSP Problem. As in Section 2, it is possible to obtain a polyhedral formulation of the RCPSFRP and adapt the *Connectivity* Theorem of Section 2.

5.2   The Insertion Financial Flow Problem

As in the case of the RCPSTDP, our main tool consists in an *Insertion* mechanism, which, at any time, considers some Timed (r^*, ϕ^*, ψ^*)-Flow $(F, F\text{-}cash, T)$ defined on some subset $W \subset V$ of the task set V and inserts some task $v_0 \in V - W$ into $(F, F\text{-}cash, T)$. So we deal with standard renewable resources $k \in K$ while considering the *Insertion Flow* model of Section 3, together with null conditional delay *Lag* values, and we introduce the *Insertion Financial Flow Problem* which is specifically related to the *Financial* resource.

The Insertion Financial Flow Problem. An *Insertion Financial Flow* instance $((X, E), z_0, A, B, In, Out, \Pi, \Pi^*, \tau, \rho, \gamma, \pi, d)$ is then defined by an almost-bipartite graph $\mathcal{N} = (X, E)$, by $z_0, A, B, In, Out, \Pi, \Pi^*$ as in Section 3, by a $\mathbb{Q}$-valued *payment delay* function $\tau \geq 0$ with domain A, and by 4 non negative coefficients d, which means the duration of task z_0, π and ρ, γ, such that: (P8)

- $\sum_{x \in A} Out_x \geq \rho;$ $\rho =$ cash amount required in order to start task z_0;
- $\sum_{x \in A} Out_x + \gamma - \rho = \sum_{y \in B} In_y;$ $\gamma =$ cash amount which is produced at the end of task z_0, and available after a delay equal to π ;

Then $G = (G_{x,y} \geq 0, x \in A \cup \{z_0\}, y \in B \cup \{z_0\}) \geq 0$, is an *Insertion Financial Flow vector* iff:

- for any x in A, $Out_x = \sum_{y \in B \cup \{x_0\}} G_{x,y}$;
- for any y in B, $In_y = \sum_{x \in A \cup \{x_0\}} G_{x,y};$
- $\gamma = \sum_{y \in B} G_{z_0,y};$ $\rho = \sum_{x \in A} G_{x,z_0};$

For such an *Insertion Financial Flow* vector G, we extend Π and τ by setting:

- $\Pi_{z_0} = Sup(Sup_{x \in A, \text{ such that } (x,z_0) \in E}(\Pi_x + d), Sup_{x \in A, \text{ such that } G_{x,z_0} \neq 0}(\Pi_x + \tau(x) + d));$ (in case no flow G_{x,z_0} exists, we set $\Pi_{z_0} = d$);
- $\tau(z_0) = \pi;$

and we define the *Makespan* value *F-Makespan(G)* by setting:

- $F\text{-}Makespan(G) = Sup(Sup_{x\in A\cup\{z_0\},y\in B \text{ such that } (x,y)\in E}(\Pi_x + \Pi_y^*),$
  $Sup_{x\in A\cup\{z_0\},y\in B \text{ such that } G_{x,y}\neq 0}(\Pi_x+\Pi_y^*+\tau(x)))$. (if no x, y exist such that
  $(x,y) \in E$ or $G_{x,y} \neq 0$, we set $F\text{-}Makespan = Sup_{y\in B}\Pi_y^*)$;

Then solving the related *Insertion Financial Flow* Problem means computing an *Insertion Financial Flow* vector G such that $F\text{-}Makespan(G)$ be minimal.

The general algorithmic *Insertion-Financial-Flow* Procedure works as the *Insertion-Flow* algorithm. Still, it behaves in some aspects in a simpler way. As a matter of fact, we notice that the case of standard resources $k \in K$ may be identified with the case of the *Insertion Financial Flow* problem when $\rho = \gamma = \pi$ and $\tau = 0$. But, for the *Insertion Financial Flow* problem, once an attachment value u in the set $\Lambda_{E,z_0} = \{u \in \mathbb{Q}$, such that:

- there does not exist $x \in A$, such that $\Pi_x > u$ and $(x, z_0) \in E$;
- there exists $x(u) \in A$, such that $\Pi_{x(u)} + \tau(x(u)) = u$;
- $\sum_{x\in A, \text{ such that } \Pi_x+\tau(x)\leq u} Out_x \geq \rho\}$

is chosen, the related *Insertion Flow* G comes through a greedy process:

- we first perform what we call the *attachment process*, which means that we compute the values $G_{x,z_0}, x \in A$ in such a way that: (P9)
  - if x and x' are such that: $G_{x',z_0} \neq 0, \Pi_x' + \tau(x') < \Pi_x + \tau(x) \leq u$, then we have: $G_{x,z_0} = Out_x$;
  - for any x such that: $\Pi_x+\tau(x) > u$ then we have: $G_{x,z_0} = 0$; This *attachment* process is completely determined by (P9) and can be performed in a greedy way. It consequently modifies the values $Out_x, x \in A$.
- then the "While Possible" loop of the *Insertion-Flow* algorithm of Section 3, which involves redirection path search, may be simplified and replaced by a greedy *Match-Flow* procedure, which implements the following lemma:

Match Flow Lemma. If the *No Cross* Property is satisfied, that means if there does not exist $x, x' \in A \cup \{z_0\}, y, y' \in B$, such that:

- $G_{x,y} \neq 0; G_{x',y'} \neq 0$
- $\Pi_{x'} + \tau(x') > \Pi_x + \tau(x); \Pi_{y'}^* > \Pi_y^*$.

then G is optimal (provided that the attachment value u has been imposed) We may state (with the same kind of proof as for Theorem 3):

Theorem 6. *The Insertion Financial Flow Problem is Time Polynomial.*

As in Section 3, we know that every time we perform the insertion process, we must apply the *Insertion Flow* and *Insertion Financial Flow* procedures respectively to every renewable resource k in the resource set K and next to the Financial resource. In order to speed the process, we make the non null $F(k)$, *F-cash* flow values be supported by the same arcs, through a *Lex-Insertion-Financial-Flow* Procedure.

5.3   Generic Flow Algorithms for the RCPSFRP Problem

The generic machinery of Section 3 gives rise to a Greedy Insertion Algorithm
RCPSFRP-Greedy-Flow, which may be randomized, and to a Local Search
RCPSFRP-LS-Flow Algorithm. Of course, both may be combined into a
GRASP algorithmic scheme. Still, there is a specific point which has to be discussed here. As told in previous sub-section A, testing the feasibility of some
RCPSFRP instance is NP-Complete. That means that, when we try the inser-
tion of some task v_0 into some into some Timed (r^*, ϕ^*, ψ^*)-Flow $(F, F\text{-}cash, T)$
defined on some subset $W \subset V$ of the task set V, there is a risk of failure, since,
for any cut U, there may be not enough cash coming from U in order to fill the
demands from both v_0 and $W - U$. As a consequence, we see that the way we
pick up tasks v of V inside the greedy **RCPSFRP-Greedy-Flow** algorithmic
scheme becomes a key issue. We call it the "Linear Ordering" component of
the RCPSFRP Problem. In order to deal with it, we associate with any RCP-
SPFRP instance $\mathcal{I} = (V, K, R, r, d, \varphi, \Phi, \phi, \psi, \delta)$, the following auxiliary problem
PLIN($\mathcal{I}$):

PLIN($\mathcal{I}$): {Compute a linear ordering σ of V such that for any task $v \in V$,
$\Phi + \sum_{w \text{ such that } w\sigma v}(\psi(w) - \phi(w)) \geq \phi(v)$}.
Then we easily notice that:

Theorem 7. *Let T be some feasible solution of the RCPSPFRP Problem in-
stance $\mathcal{I} = (V, K, R, r, d, \varphi, \Phi, \phi, \psi, \delta)$. Then, any linear ordering σ which is compatible with T, d and δ (that means such that if $T(v) + d(v) + \delta(v) \leq T(w)$ then
$v \sigma w$) is a solution of PLIN($\mathcal{I}$)). Conversely, if we apply the **RCPSFRP-
Greedy-Flow** algorithm to $\mathcal{I}$ while picking up the tasks of V according to a
feasible solution σ of PLIN($\mathcal{I}$), then we get a feasible solution of $\mathcal{I}$.*

So, we handle the Linear Ordering component of the RCPSFRP Problem
while designing the **RCPSFRP-Greedy-Flow** algorithmic scheme:

- Compute a set Λ of N distinct solutions of PLIN($\mathcal{I}$), (N = replication parameter); we try to get solutions which are not too close to each other;
- Then, *For* any $\sigma \in \Lambda$, we apply the **RCPSFRP-Greedy-Flow** algorithm
  while picking up the tasks v of V according to the linear ordering σ;
- Keep the best solution $T = T(\sigma)$, $\sigma \in \Lambda$, produced by the *For* loop.

We adapt the **RCPSFRP-LS-Flow** algorithmic scheme by the same way:

- Starting from a current feasible solution T of the RCPSFRP instance $\mathcal{I}$, we
  first compute a linear ordering σ which is compatible with T, d and δ ;
- Next, every time we perform the removal of some subset W of V, we compute
  a set Λ of M distinct linear orderings σ^* of V, which coincide with σ on $V - W$
  and which are solution of PLIN($\mathcal{I}$), M being some parameter value;
- We denote by Λ_W the set of the restrictions of σ, $\sigma \in \Lambda$ to W;
- For every linear ordering τ in Λ_W, we try the reinsertion of W while picking
  up the tasks of W according to τ (Failure may occur).

5.4   Numerical Experiments

We performed the same kind of experiments as for the RCPSTLP Problem. We provide here results related the behaviour of the **RCPSFRP-Greedy-Flow** Procedure. We used 30 and 60 task instances I from the PSPLIB library, together with some solution T(I) obtained from the tests of Section IV. For every such an instance I, we introduced an additional Financial resource and generated values $\phi(v)$, $\varphi(v)$, $\delta(v)$, $v \in V$, and Φ, in such a way that, for every instance I, T(I) remained a feasible solution with unchanged makespan MK(I). We denoted by Gap-FR the gap (in %) between the value obtained by the **RCPSFRP-Greedy-Flow Procedure** and the almost-optimal value MK(I), and we got results according to the table.

Table 3. *RCPSP-Greedy-Flow* behaviour, Mean Results

N-Ac	N-res	N-re	Time(s)	Gap-FR
30	4	10	0.05	9.7
30	4	1000	5.3	3.7
60	4	10	0.61	13.8
60	4	1000	73.7	4.8

Comment: one handling Financial resource is more difficult. The Linear Ordering issue is critical.

6   Scheduling through Borrow/Invest Strategies

We suppose now that, at any time during the execution of the project, we are able to borrow money at a constant rate ρ_B and to invest money at a constant rate ρ_I. An *Investment plan* is a triple (t, Δ, M): M money is invested at time t, which gives back $M.(1 + \rho_I).\Delta$ money at time $t + \Delta$. *Invest* is the set of all *Investment* plans. By the same way a *Loan plan* is a triple (t, Δ, M): M money is borrowed at time t, and $M.(1 + \rho_B).\Delta$ money is given back at time $t + \Delta$. We denote by *Loan* the set of all *Loan* plans.

Let $\mathcal{I} = (V, K, R, r, d, \varphi, \Phi, \phi, \psi, \delta)$ be some RCPSFRP instance, and $\mathcal{I}(\Psi)$ be the instance which derives from $\mathcal{I}$ by replacing Φ by $\Phi + \Psi$. For a given value Ψ, let $S = (F, F\text{-}Cash, T)$ be some solution of $\mathcal{I}(\Psi)$. At any time t in $[0, T_{End}]$, this solution provides us with an available money amount $Cash_S(t)$ (the quantity $\mathcal{M}(T, t)$ of section 5). The function $t \rightarrow Cash_S(t) - \Psi$ may take negative values, which forbids $(F, F\text{-}Cash, T)$ from being a feasible solution of $\mathcal{I} = \mathcal{I}(0)$. Still, we may use convenient sets $Inv \subset Invest$ and $Lo \subset Loan$ in order to *balance* the $t \rightarrow Cash_S(t) - \Psi$ function, that means to make that, at any time t, $Cash_S(t) - \Psi + Inv_0(t) + Lo_1(t) - Inv_1(t) - Lo_0(t) \geq 0$, where:

- $Inv_0(t) =$ Money produced by the plans in *Inv* which end at time $\leq t$;
- $Inv_1(t) =$ Money required by the plans in *Inv* which start at time $\leq t$;
- $Lo_0(t) =$ Money given back by plans in *Lo* which end no later than t;
- $Lo_1(t) =$ Money obtained from the plans in *Lo* which start no later than t.

6.1   The 2nd RCPSFRP Problem

Thus, solving the *Second Resource Constraint Project Scheduling Problem with Financial Resource* defined by $\mathcal{I} = (V, K, R, r, d, \varphi, \Phi, \phi, \psi, \delta)$ and by the interest rates ρ_B and ρ_I, means computing: Ψ, $(F, F\text{-}Cash, T)$, Inv and Lo in such a way the $t \rightarrow Cash_S(t) - \Psi$ *Available Cash* function be *balanced*, and T_{End} be minimal.

Example. As in the previous section, with ρ_B: 50%, ρ_I: 20%, $\delta = 0$, $\Phi = 5$. Figure 4 shows the *Available Cash* function: Balancing strategy: First *Loan* plan: (0, 2, 2); Second *Loan* plan: (2, 1, 7.5); Last *Investment* plan: (3, 4, 2.25).

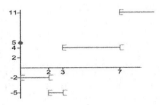

Fig. 4. The *Available Cash* function with Initial Cash Amount = 5

6.2   Handling the 2nd RCPSFRP Problem

We call *Balancing* problem, the problem about the search of sets $Inv \subset Invest$ and $Lo \subset Loan$ which *balance* the $t \rightarrow Cash_S(t) - \Psi$ function. It can be handled through a **Balancing** Procedure which implements the following result:

Theorem 8. *An optimal Inv and Loan strategy may be computed as the one which makes the Available Cash null during the whole interval execution* $[0, T_{End}]$. *It succeeds in Balancing the* $t \rightarrow Cash_S(t) - \Psi$ *function if the Available Cash at* T_{End} *is* ≥ 0.

Thus, we handle the 2nd RCRCFRP Problem through the dichotomic scheme:

Algorithm 2: Second-Financial-Flow Algorithm

Input: $\mathcal{I} = (V, K, R, r, d, \varphi, \Phi, \phi, \psi, \delta)$, ρ_I, ρ_B;
Output: Ψ, a solution $(F, F\text{-}Cash, T)$ of $\mathcal{I}(\Psi)$, plans $Lo(\Psi)$, $Inv(\Psi)$;
$\Psi_{min} \leftarrow 0$; $(F_{min}, F\text{-}Cash_{min}, T_{min}) \leftarrow I(\Psi_{min})$;
$\Psi_{max} \leftarrow \sum_{v \in V} \phi(v) - \Phi$; $(F_{max}, F\text{-}Cash_{max}, T_{max}) \leftarrow$ solution of $I(\Psi_{max})$;
if *the* Balancing *Problem related to* Ψ_{max} *has a solution* **then** $\Psi \leftarrow \Psi_{max}$
else
$\quad$ Not Stop;
$\quad$ **while** *Not stop* **do**
$\quad\quad$ $\Psi_{med} \leftarrow (\Psi_{max} + \Psi_{min})/2$;
$\quad\quad$ **if** *the Balancing Problem related to* Ψ_{med} *admits a solution* Inv,
$\quad\quad$ Lo **then** $\Psi_{min} \leftarrow \Psi_{med}$ **else** $\Psi_{max} \leftarrow \Psi_{med}$;
$\quad\quad$ **if** $\Psi_{max} - \Psi_{min}$ *is small enough* **then** Stop; $\Psi \leftarrow \Psi_{min}$

We focused here on the impact of Borrow/Invest strategies. We used 30 task PSPLIB like feasible instances. For every instance, we computed the gap GAP-F = (V-1 V-Standard)/V-Standard between the standard RCPSP optimal value V-Standard and the value V-1 obtained through RCPSFRP-Greedy-Flow, and the gap GAP-Rate = (V-1 V-Rate)/V-Rate between V-1 and the value V-Rate derived from Second-Financial-Flow. We got:

Table 1. Impact of Borrow/Invest strategies: Procedure Second-Financial-Flow

ID	rI	rB	GAP-F (%)	Gap-Rate (%)
1	0.1	0.2	73	67
2	0.3	0.5	24	15
3	0.5	0.7	5.2	35
4	0.3	0.3	98	28
5	0.5	0.5	19	19

Comment: in case Financial constraint significantly slows down the project, Borrow/Invest strategies ease the effect of this Financial constraint.

7   Conclusion

Our RCPSP flow models rely on very generic features, which allow fast software implementation. Several questions should be studied, from both theoretical and practical point of view, related to the way *Cuts* are generated and about what we called the *Linear Ordering* component of our algorithms.

References

1. Kolisch, R., Padman, R.: Deterministic project scheduling. Omega 48, 249–272 (1999)
2. Brucker, P., Drexl, A., Mohring, R., Neumann, K., Pesch, E.: RCPSP: notation, classification, models and methods. EJOR 112, 3–41 (1999)
3. Baptiste, P., Laborie, P., Lepape, C., Nuijten, W.: Constraint-based scheduling/planning. In: Rossi, F., Van Beek, P. (eds.) Handbook Constraint Prog., ch. 22, pp. 759–98. Elsevier (2006)
4. Sauer, N., Stone, M.G.: Rational preemptive scheduling. Order 4, 195–206 (1987)
5. Blazewiecz, J., Ecker, K.H., Schmlidt, G., Weglarcz, J.: Scheduling in computer and manufacturing systems, 2nd edn. Springer, Berlin (1993)
6. Herroelen, W.: Project Scheduling-Th./Pract. Prod./Op. Manag. 14(4), 413–432 (2006)
7. Liu, S.S., Wang, C.J.: RCPSP profit max with cash flow. Aut. Const. 17, 966–974 (2008)
8. Damay, J., Quilliot, A., Sanlaville, E.: Linear programming based algorithms for preemptive and non preemptive RCPSP. EJOR 182(3), 1012–1022 (2007)

9. Moukrim, A., Quilliot, A.: Preemptive scheduling on parallel machines. O.R Let. 33, 143–151 (2005)

10. De Reyck, B., Herroelen, W.: Branch/bound for the RCPSP with generalized precedence relations. EJOR 111, 152–174 (1998)

11. Mohring, R.H., Rademacher, F.J.: Scheduling problems with resource duration interactions. Methods of Operat. Research 48, 423–452 (1984)

12. Kimms, A.: Mathematical programming and financial objectives for scheduling projects, OR and Management Sciences. Kluwer Academic Publisher (2001)

13. Chtourou, H., Haouari, M.: A two-stage-priority rule based algorithm for robust resource-constrained project scheduling. Computers and Ind. Eng., 12 pages (2008)

14. Haouari, M., Gharbi, A.: A improved max-flow based lower bound for minimizing maximum lateness on identical parallel machines. OR Letters 31, 49–52 (2003)

15. Carlier, J., Neron, E.: Computing redundant resources for the resource constrained project scheduling problem. EJOR 176, 1452–1463 (2007)

16. Hartmann, S., Briskorn, D.: A survey of variants of RCPSP. EJOR 207, 1–14 (2010)

17. Kolisch, R., Hartmann, S.: Heuristic for RCPSP: computational analysis. In: Weglarcz, J. (ed.) Project Scheduling: Models and Applications. Kluwer Press (1999)

18. Demeulemeester, E., Herroelen, W.: New benchmark for mult. RCPSP. Management Sciences 43, 1485–1492 (1997)

19. Brucker, P., Knust, S., Schoo, A., Thiele, O.: A branch and bound algorithm for the resource constrained project scheduling problem. EJOR 107, 272–288 (1998)

20. Mingozzi, A., Maniezzo, V., Ricciardelli, S., Bianco, L.: An exact algorithm for RCPSP based on a new math. formulation. Manag. Sc. 44, 714–729 (1998)

21. Baptiste, P., Demassey, S.: Tight LP-bounds for RCPSP. OR Spect. 2, 251–262 (2004)

22. Brucker, P., Knust, S.: A linear programming and constraint propagation based lower bound for the RCPSP. EJOR 127, 355–362 (2000)

23. Artigues, C., Roubellat, F.: A polynomial activity insertion algorithm in a multiresource schedule with cumulative constraints. EJOR 127(2), 297–316 (2000)

24. Artigues, C., Michelon, P., Reusser, S.: Insertion for static/dyn. RCPSP. EJOR 149, 249–267 (2003)

25. Palpant, M., Artigues, C., Michelon, P.: LSSPER: solving RCPSP with large neighbourhood search. Annals of O.R 131(1-4), 237–257 (2004)

26. Kolisch, R., Hartmann, S.: Experimental investigation of heuristics for the resource con-strained scheduling problem: an update. EJOR 174, 23–37 (2006)

27. Ahuja, R.V., Magnanti, T.L., Orlin, J.B.: Network Flows: Theory/Appl., Prentice, N.J (1993)

A New Hybrid GA-FA Tuning of PID Controller for Glucose Concentration Control

Olympia Roeva and Tsonyo Slavov

Institute of Biophysics and Biomedical Engineering, BAS
105 Acad. G. Bonchev Str., 1113 Sofia, Bulgaria
Technical University of Sofia
8 Kliment Ohridski Bulv., 1000 Sofia, Bulgaria
olympia@biomed.bas.bg,
ts_slavov@tu-sofia.bg

Abstract. In this paper a hybrid scheme using Firefly Algorithm (FA) - Genetic Algorithm (GA) is introduced. The novel hybrid meta-heuristics algorithm is realized and applied to PID controller parameter tuning in Smith Predictor for a nonlinear control system. The controller is used to control feed rate and to maintain glucose concentration at the desired set point for an *E. coli* MC4110 fed-batch cultivation process. The hybrid FA-GA adjustments are done based on several pre-tests. Simulation results indicate that the applied hybrid algorithm is effective. Good closed-loop system performance is achieved on the basis of the considered PID controllers tuning procedures. Moreover, the observed results are compared to the ones obtained by applying the pure FA and pure GA. The comparison shows that the proposed hybrid algorithm is highly competitive with standard FA and GA for considered here optimization problem.

Keywords: meta-heuristics, firefly algorithm, genetic algorithm, *E. coli* cultivation process, PID controller, parameter tuning.

1 Introduction

Among the different modes of operation concerning cultivation processes, (batch, fed-batch and continuous), fed-batch operation is the most often used one in industry. Since both nutrient overfeeding and underfeeding is detrimental to cell growth and product formation, development of a suitable feeding strategy control is critical in fed-batch cultivation processes. The control strategy for substrate feed rate can be summarized in three groups: open (feedforward), closed-loop (feedback) control and mixed (feedforward-feedback). In feedback control of industrial cultivation processes, the proportional-integral-derivative (PID) controller is widely used [4], [10].

Usually the PID controller is poorly tuned due to highly changing dynamics of most bioprocesses caused by the nonlinear growth of the cells and the changes

S. Fidanova (Ed.): *Recent Advances in Computational Optimization*, SCI 470, pp. 155–168.
DOI: 10.1007/978-3-319-00410-5_9   © Springer International Publishing Switzerland 2013

in the overall metabolism. The tuning procedure is a significant challenge for the conventional optimization methods. As an alternative, meta-heuristics could be applied [12], [14], [15].

During the last decade, a broad class of meta-heuristics has been developed and applied to a variety of areas. Algorithms like Genetic Algorithms (GA) and evolution strategies, ant colony optimization, artificial bee colony optimization, bacterial foraging algorithms, particle swarm optimization, tabu search, simulated annealing, multi-start and iterated local search are - among others - often listed as examples of classical meta-heuristics, and they have individual historical backgrounds and follow different paradigms and philosophies [6], [7], [24], [25]. Recently, a new meta-heuristics called Firefly Algorithm (FA) has emerged. This algorithm was proposed by Xin-She Yang [28]. The FA is very efficient and can outperform other meta-heuristics, such as genetic algorithms, in solving many optimization problems [20], [29], [30].

In the literature, there are results showing different strategies based on meta-heuristic algorithms for the optimal tuning of PID controllers considering only linear systems [20]. Actually, there is a lack of results about using meta-heuristic algorithms for bioprocess control design, considering nonlinear systems.

Hybrid algorithms have received significant interest in recent years and are being increasingly used to solve real-world problems [1]. Usually different local search methods have got attention in such combinations [8], [16], [26].

In this paper a hybrid meta-heuristic algorithm FA-GA is introduced for the first time specified to solve PID controller parameter tuning. The idea is to combine the two meta-heuristics, namely FA in order to explore the search place either to isolate the most promising region of the search space and GA - to exploit the information gathered by the FA. An optimization algorithm based on FA-GA hybrid is realized and applied for parameter tuning of the PID Controller for glucose concentration control of a nonlinear *E. coli* fed-batch cultivation process.

2   Problem Formulation

A modified Smith Predictor (SP) structure, proposed in [17], based on a nonlinear plant model is used here. When the object is characterized with a significant time delay, the conventional PID controller can not ensure the control system performance. A tool approved in the practice for time delay compensation is the SP [23]. In this predictor scheme, the mathematical model of the "nondelayed" process is implemented in an internal feedback loop around a conventional controller. The major advantage of the SP is that the delay issues can be ignored in controller design [9].

The structure of the control system is shown in Fig. 1.

In the conventional case of SP, only the predicted by model output is used to form the inner feedback. In this case, the controller uses plant 's output and the process variables both predicted by nonlinear process model [17]. They are used to form the feedback and the feedforward terms of the control signal, respectively. The feedforward term is utilized to hold the nonlinear plant at the actual equilibrium point.

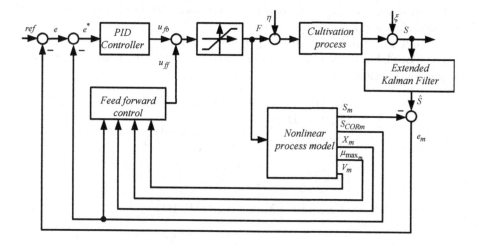

Fig. 1. Structure of the control system

The mathematical model of the considered process (block labeled "Cultivation process") can be represented by [22]:

$$\begin{vmatrix} \dot{\mathbf{x}}(t) = f(\mathbf{x}, F) + \eta(t) \\ S(t) = \mathbf{H}\mathbf{x}(t) + \xi(t) \end{vmatrix} \tag{1}$$

$$\mathbf{x}(t) = \begin{bmatrix} X(t) & S(t) & V(t) & \mu_{max}(t) \end{bmatrix}^{\mathrm{T}} \tag{2}$$

$$f(\mathbf{x}, F) = \begin{bmatrix} \mu_{max}(t) \dfrac{S(t)}{k_S + S(t)} X(t) - \dfrac{F(t)}{V(t)} X(t) \\ -\dfrac{1}{Y_{S/X}} \mu_{max}(t) \dfrac{S(t)}{k_S + S(t)} X(t) + \dfrac{F(t)}{V(t)} (S_{in} - S(t)) \\ F(t) \\ 0 \end{bmatrix} \tag{3}$$

$$\mathbf{H} = \begin{bmatrix} 0 & 1 & 0 & 0 \end{bmatrix} \tag{4}$$

$$\eta(t) = \begin{bmatrix} \eta_X(t) & \eta_S(t) & 0 & \eta_{\mu_{max}}(t) \end{bmatrix}^{\mathrm{T}} \tag{5}$$

where: $\mathbf{x}$ is the state vector; f is nonlinear model function; $\eta(t)$ is process noise; $\mathbf{H}$ is output matrix; $\xi(t)$ is measurement noise, $[\mathrm{g\cdot l^{-1}}]$; X is concentration of biomass, $[\mathrm{g\cdot l^{-1}}]$; S is concentration of substrate (glucose), $[\mathrm{g\cdot l^{-1}}]$; F is feed rate, $[\mathrm{l\cdot h^{-1}}]$; S_{in} is substrate concentration of the feeding solution, $[\mathrm{g\cdot l^{-1}}]$; V is bioreactor volume, $[\mathrm{l}]$; μ_{max} is maximum growth rate, $[\mathrm{h^{-1}}]$; k_S is saturation constant, $[\mathrm{g\cdot^{-1}}]$; $Y_{S/X}$ is yield coefficient, $[\text{-}]$; η_X is biomass concentration process noise, $[\mathrm{g^2\cdot l^{-2}\cdot h^{-2}}]$; η_S is substrate concentration process noise, $[\mathrm{g^2\cdot l^{-2}\cdot h^{-2}}]$; $\eta_{\mu_{max}}$ is the maximum growth rate process noise, $[\mathrm{h^{-1}}]$.

The model inaccuracy is modeled via zero mean white Gausian noise. The corresponding variances are [3]:
$\eta_X = 0.001$ g$^2 \cdot$l$^{-2} \cdot$h^{-2}, $\eta_S = 0.001$ g$^2 \cdot$l$^{-2} \cdot$h^{-2} and $\eta_{\mu_{\max}} = 0.05$ l$\cdot$h^{-3}.

The block labeled "Nonlinear process model" predicts the non-delayed model output by equations:

$$
\begin{vmatrix}
\dot{\mathbf{x}}_m(t) = f_m(\mathbf{x}_m, F) \\
S_m(t) = \mathbf{H}_m \mathbf{x}_m(t) \\
S_{CORm}(t) = S_m(t) + \dfrac{\mu_m(t)\, X_m(t)}{Y_{S/X}} \Delta t, \\
\mu_m(t) = \mu_{\max_m} \dfrac{S_m(t)}{k_S + S_m(t)}
\end{vmatrix}
\tag{6}
$$

$$
\mathbf{x}_m(t) = \begin{bmatrix} X_m(t) & S_m(t) & V_m(t) \end{bmatrix}^{\mathrm{T}}
\tag{7}
$$

$$
f_m(\mathbf{x}_m, F) = \begin{bmatrix}
\mu_m(t)\, X_m(t) - \dfrac{F(t)}{V_m(t)} X_m(t) \\
-\dfrac{1}{Y_{S/X}} \mu_m(t)\, X_m(t) + \dfrac{F(t)}{V_m(t)} (S_{in} - S_m(t)) \\
F(t)
\end{bmatrix}
\tag{8}
$$

$$
\mathbf{H}_m = \begin{bmatrix} 0 & 1 & 0 \end{bmatrix}
\tag{9}
$$

where: $\mathbf{x}_m$ is the state vector; f_m is nonlinear model function; $\mathbf{H}_m$ is output matrix; X_m is the evaluated by model concentration of biomass, [g$\cdot$l^{-1}]; S_m is delayed concentration of substrate (glucose) evaluated by model, [g$\cdot$l^{-1}]; S_{CORm} is non-delayed concentration of substrate predicted by model, [g$\cdot$l^{-1}]; V_m is evaluated by model bioreactor volume, [l]; μ_m is specific growth rate, [h^{-1}]. Here $\mu_{max_m} = 0.5$ h^{-1}.

To obtain the glucose concentration estimate an Extended Kalman filter (EKF) is designed [22]. Based on discretization of process model (Eq. (1)) the following EKF is obtained [22]:

$$
\begin{vmatrix}
\hat{\mathbf{x}}(k+1) = \mathbf{f_d}(\hat{\mathbf{x}}(k)) + \mathbf{K_{EKF}}(k+1)\, (S(k+1) - \mathbf{H} f_d(\hat{x}(k))) \\
\hat{S}(k+1) = \mathbf{H}\hat{\mathbf{x}}(k+1)
\end{vmatrix}
\tag{10}
$$

$$
\hat{\mathbf{x}}(0) = \begin{bmatrix} 1.25 & 0.8 & 1.35 & 0.5 \end{bmatrix}^{\mathrm{T}}
\tag{11}
$$

$$
\mathbf{f_d}(\hat{\mathbf{x}}(k)) = \hat{x}(k) + T_0 \mathbf{f}(\hat{\mathbf{x}}(k))
\tag{12}
$$

where: $\hat{\mathbf{x}}(\cdot)$ and $\hat{S}(\cdot)$ are the estimates of $\mathbf{x}(\cdot)$ and $S(\cdot)$; $\mathbf{K_{EKF}}(\cdot)$ – the EKF gain.

The PID controller algorithm is described as follows:

$$
u_{fb}(s) = K_p\big(b\,e(s) - S_{CORm}(s)\big) + \frac{K_p}{T_i s} e^*(s) +
$$
$$
+ \frac{T_d s}{1 + \dfrac{T_d s}{N}} \big(c\,e(s) - S_{CORm}(s)\big)
\tag{13}
$$

where: $u_{fb}(s)$ is the feedback term of control variable, $[\mathrm{l} \cdot \mathrm{h}^{-1}]$; K_p is proportional gain, [-]; T_i is integral time, [h]; T_d is derivative time, [h]; b and c are set-point weight coefficients, [-]; T_d/N is low-pass first order filter of D-term time-constant, [h].

The error $e^*(s)$ is:

$$e^*(s) = e(s) - S_{CORm}(s),$$

where:

$e(s) = ref(s) - e_m(s)$, $ref(s)$ is a reference signal and $e_m(s) = \hat{S}(s) - S_m(s)$.

For the *E. coli* MC4110 cultivation considered here, the process desired set-point (reference signal) is set at $ref(s) = 0.1 \ \mathrm{g} \cdot \mathrm{l}^{-1}$ glucose concentration [3].

Considering real applications, usually a digital PID controller is implemented. Here, for discretization of the PID controller (Eq. (13)), the backward Euler method [20] is used. The mathematical description of the designed digital PID controller is:

$$u_{fb}(k) = u_p(k) + u_i(k) + u_d(k) \tag{14}$$

$$u_p(k) = K_p\big(b\,e(k) - S_{CORm}(k)\big) \tag{15}$$

$$
\begin{aligned}
u_i(k) = u_i(k-1)+ \\
+b_{i1}\big(e(k) - S_{CORm}(k)\big) + b_{i2}\big(e(k-1) - S_{CORm}(k-1)\big)
\end{aligned}
\tag{16}
$$

$$
\begin{aligned}
u_d(k) = a_d u_d(k-1)+ \\
+b_d\big(ce(k) - ce(k-1) - S_{CORm}(k) + S_{CORm}(k-1)\big)
\end{aligned}
\tag{17}
$$

where

$$b_{i1} = K_p\frac{T_0}{T_i}; \ \ b_{i2} = 0; \ \ a_d = \frac{T_d}{T_d + NT_0}; \ \ b_d = K_p\frac{T_dN}{T_d + NT_0}.$$

The control variable used to control the feed rate is:

$$F(k) = u_{fb}(k) + u_{ff}(k) \tag{18}$$

where

$$u_{ff}(k) = \frac{1}{Y_{S/X}} \frac{V_m(k)\mu_m(k)X_m(k)}{S_{in} - S_{CORm}} \tag{19}$$

is a feedforward term obtained from the steady state conditions.

To provide control action designed for specific process requirements, tuning of the PID controller parameters is required. The controller parameters are K_p, T_i, T_d, b, c and N.

3   Hybrid Firefly Algorithm - Genetic Algorithm

The proposed FA-GA algorithm is basically a combination of the FA and GA methods. In this hybrid, in the first step, FA explores the search place in order to either isolate the most promising region of the search space. In the second step, to improve global search and get rid of trapping into several local optima, it is introduced GA to explore search space and find new better solutions. Below is presented a brief description of the FA and GA techniques.

3.1   Firefly Algorithm

The FA is a meta-heuristic algorithm which is inspired from flashing light behaviour of fireflies in nature. The pattern of flashes is often unique for a particular species of fireflies. The two basic functions of such flashes are to attract mating partners or communicate with them, and to attract potential victim. Additionally, flashing may also serve as a protective warning mechanism.

In FA, each firefly has a location $y = (y_1, ..., y_d)^{\mathrm{T}}$ in a d-dimensional space and light intensity $I(y)$ or attractiveness $\beta(y)$, which are proportional to an objective function $f(y)$. Attractiveness $\beta(y)$ and light intensity $I(yx)$ are relative and these should be judged by the rest fireflies. Thus, attractiveness will vary with the distance $r_{i,j}$ between firefly i and firefly j. So, attractiveness β of a firefly can be defined by Eq. (20) ([28], [29], [30]):

$$\beta(r) = \beta_0 e^{-\gamma r^m}, m \geq 1 \tag{20}$$

where r (or $r_{i,j}$) is the distance between the i-th and j-th of two fireflies. β_0 is the initial attractiveness at $r = 0$ and γ is a fixed light absorption coefficient that controls the decrease of the light intensity. In the herewith applied FA, the coefficient $m = 2$.

The initial solution is generated based on:

$$y_j = rand(Ub - Lb) + Lb \tag{21}$$

where $rand$ is a random number generator uniformly distributed in the space $[0, 1]$; Ub and Lb are the upper range and lower range of the j-th firefly, respectively. When firefly i is attracted to another more attractive firefly j, its movement is determined by:

$$y_{i+1} = y_i + \beta_0 e^{-\gamma r_{i,j}^2}(y_i - y_j) + \alpha(rand - \frac{1}{2}) \tag{22}$$

where the first term is the current position of a firefly, the second term is used for considering a firefly's attractiveness to light intensity seen by adjacent fireflies $\beta(r)$ (Eq. (20)), and the third term is used to describe the random movement of a firefly in case there are no brighter ones. The coefficient α is a randomization parameter determined by the problem of interest. The distance $r_{i,j}$ between any two fireflies i and j at y_i and y_j, respectively, is defined according to [28], [29], [30]:

$$r_{i,j} = \|y_i - y_j\| = \sqrt{\sum_{k=1}^{d} (y_{i,k} - y_{j,k})^2} \tag{23}$$

where $y_{i,k}$ is the k-th component of the spatial coordinate y_i of the i-th firefly.

3.2   Genetic Algorithm

GA originated from the studies of cellular automata, conducted by John Holland and his colleagues at the University of Michigan. Holland's book [13], published in 1975, is generally acknowledged as the beginning of the research of genetic algorithms. Since their introduction and subsequent popularization [13], the GA have been frequently used as an alternative optimization tool to the conventional methods [11] and have been successfully applied in a variety of areas, and still find increasing acceptance [1], [2], [5], [8], [16], [19], [21].

The GA is a model of machine learning which derives its behavior from a metaphor of the processes of evolution in nature [11]. This is done by the creation within a machine of a population of individuals represented by chromosomes. A chromosome could be an array of real numbers, a binary string, a list of components in a database, all depending on the specific problem.

The GA maintains a population of individuals, $P(t) = y_1^t, ..., y_n^t$ for generation t. Each individual represents a potential solution to the problem and is implemented as some data structure U. Each solution is evaluated to give some measure of its "fitness". Fitness of an individual is assigned proportionally to the value of the objective function of the individuals. Then, a new population (generation $t+1$) is formed by selecting more fit individuals (selected step). Some members of the new population undergo transformations by means of "genetic" operators to form new solution.

There are unary transformations m_i (mutation type), which create new individuals by a small change in a single individual ($m_i : U \rightarrow U$), and higher order transformations c_j (crossover type), which create new individuals by combining parts from several individuals ($c_j : U \times ... \times U \rightarrow U$).

After some number of generations the algorithm converges - it is expected that the best individual represents a near-optimum (reasonable) solution.

The combined effect of selection, crossover and mutation gives so-called reproductive scheme growth equation [11]:

$$\xi(U, t+1) \geq \xi(U, t) \, eval(U, t) / \bar{F}(t) \left[1 - p_c \frac{\delta(U)}{m-1} - o(U) \, p_m \right] \qquad (24)$$

The structure of the hybrid FA-GA is shown by the pseudo-code in Fig. 2.

4   Results and Discussion

A series of tuning procedures for the considered control system using hybrid FA-GA, pure GA and pure FA are performed. Computer specifications to run all optimization procedures are Intel Core i5-2320 CPU 3.00GHz, 8 GB Memory (RAM), Windows 7 (64bit) operating system.

The main FA parameters are set to the optimal settings [20]:

- $\beta_0 = 1$, $\gamma = 1$, $\alpha = 0.2$,
- number of fireflies = 25, number of iterations = 50.

```
begin FA
Define
    algorithm parameters and operators
    objective function f(y), where y = (y₁, ..., y_d)ᵀ
Generate initial population of fireflies yᵢ, (i = 1, 2, ..., n)
Determine light intensity Iᵢ at yᵢ via f(yᵢ)
    while (t < FA_MaxIteration) do
        for i = 1 : n all n fireflies do
            for j = 1 : i all n fireflies do
                if (I_j > I_i) then
                Move firefly i towards j
                end if
                Attractiveness varies with distance r via exp[−γr²]
                Evaluate new solutions and update light intensity
            end for j
        end for i
        Rank the fireflies and find the current best
    end while
    Final best population of fireflies
end begin FA
begin GA
    i = 0
    Initial population P(0) = Final best population of fireflies
    Evaluate P(0) fitness
        while (t < GA_MaxGeneration) do
            i = i + 1
            Select P(i) from P(i − 1)
            Recombine P(i) with crossover probability p_c
            Mutate P(i) with mutation probability p_m
            Evaluate P(i) fitness
        end while
        Rank the chromosomes, find the current best and save
        Postprocess results and visualization
end begin GA
```

Fig. 2. Pseudo-code for hybrid FA-GA

The main GA parameters are as follows [20]:

- number of individuals = 25; number of generations = 50;
- double point crossover with crossover probability $p_c = 0.7$;
- mutation with low probability $p_m = 0.01$;
- a roulette wheel mechanism is employed;
- a generation gap of 0.97 is chosen,
- fitness-based reinsertion is used.

For realistic comparison, the pure GA and FA are run for the same number of function evaluations, namely 1250.

Some of the hybrid FA-GA parameters are tuned based on several pre-tests according to the problem considered here. As a result the following parameters are used:

- number of fireflies = 15,
- number of FA iterations = 10,
- number of GA individuals = 15,
- number of GA generations = 20.

The rest of the hybrid algorithm parameters are the same as the above listed algorithm parameters for GA and FA. Because of the stochastic characteristics of the applied algorithms, a series of 30 runs for each algorithm are performed and the best, the worst and average numerical results are obtained and presented here.

The range of the tuning parameters is considered, as follows:
$K_p \in [0, inf]$; $T_i \in [0, inf]$; $T_d \in [0, inf]$; $b, c \in [0, inf]$; $N \in [0, inf]$.

To evaluate the significance of the tuning procedure the integrated square error (I_{ISE}) criterion is used:

$$I_{ISE} = \int_0^T e^*(t)^2 \mathrm{d}t \tag{25}$$

where t is time, h; T is end time of the cultivation, h.

As a result of the tuning procedures, the optimal PID controllers settings are obtained. Thus, for a short time, the controllers set the control variable and maintain it at the desired set point ($ref(s) = 0.1$ g·l^{-1}) to the end of fed-batch cultivation process.

The numerical values of the controllers parameters (K_p, T_i, T_d, b, c and N), objective functions values (I_{ISE}), total computational times (T_{comp}) and number of functions evaluations (N_{FE}) are presented in Table 1.

The presented experimental results show that the proposed hybrid FA-GA has superior performance compared to the both pure FA and GA tuned PID controllers. These results are obtained for the "best" and "average" values.

For about two times less number of function evaluations (from $N_{FE} = 1250$ to $N_{FE} = 670$) and for about 27% less computational time (see Table 1), the hybrid FA-GA achieves the values:

- best values: $I_{ISE}^{FA\text{-}GA} = 16.8400$;
- average values: $I_{ISE}^{FA\text{-}GA} = 16.8720$.

The results considering the "worst" values are achieved from the FA tuned PID controller. In the other two cases ("best" and "average") for the same number of function evaluations ($N_{FE} = 1250$) the FA tuned PID controller keeps the glucose concentration at the desired set point more accurately than the GA tuned PID controller:

- best values: $I_{ISE}^{FA} = 16.8410$ vs. $I_{ISE}^{GA} = 16.8706$;
- average values: $I_{ISE}^{FA} = 16.8720$ vs. $I_{ISE}^{GA} = 16.8823$.

Table 1. Experimental results of PID controller tuning for glucose concentration control during an *E. coli* MC4110 fed-batch cultivation process

Value	Algorithm	N_{FE}	T_{comp}	I_{ISE}	PID Controller Parameters					
					K_p	T_i	T_d	b	c	N
best	GA	1250	185.8908	16.8706	0.522	0.198	0.006	1.037	0.976	1.236
	FA	1250	178.7211	16.8410	0.303	0.491	0.003	0.099	1.621	19.952
	FA-GA	**670**	**131.2417**	**16.8400**	**0.684**	**0.526**	**0.002**	**0.834**	**1.235**	**17.221**
worst	GA	1250	184.6116	16.9548	0.413	0.005	0.008	0.566	0.830	1.470
	FA	**1250**	**181.0312**	**17.0793**	**0.256**	**0.609**	**0.010**	**1.545**	**0.835**	**10.180**
	FA-GA	670	138.3622	17.0312	0.029	0.560	0.045	2.340	0.666	14.876
average	GA	1250	178.3081	16.8823	0.135	0.948	0.024	2.371	0.834	1.433
	FA	1250	174.7211	16.8842	0.388	0.521	0.005	0.631	1.207	13.496
	FA-GA	**670**	**137.8752**	**16.8720**	**0.847**	**0.210**	**0.003**	**0.997**	**0.972**	**16.415**

Although the observed best objective function values of the hybrid FA-GA and FA are very close, the FA-GA tuned PID controller shows better performance than the FA tuned one. In the next figures, some graphical results of the control system performance for *E. coli* fed-batch cultivation process are presented.

In Fig. 3 obtained profiles of the control variable (glucose concentration) for the three PID controllers are shown. The cultivation process is simulated for 16 hours that is with 3 hours more than the process discussed in [3]. Thus, it is shown that the resulting PID controllers ensured good control system performance for a much longer cultivation time. As it can be seen from Fig. 3, for up to 15 h the three controllers (FA, GA and hybrid FA-GA) show identical performance with respect to the resulting dynamics of control variable. In the same time the errors between control variable and reference signal (I_{ISE}) are identical (see Table 1). After 15 h, the errors of control systems based on both GA and FA tuned controllers are greather than the corresponding value for the control system based on the controller tuned by proposed here hybrid FA-GA (Fig. 3, solid bold line).

On Fig. 4 the control signal (feed rate profiles) for the three PID controllers are presented. The figure shows analogical results. The control signal of the FA-GA PID controller is smoother than the corresponding profiles calculated by standard GA and FA conrollers. This result is more expressive in the last two hours of *E. coli* fed-batch cultivation process.

The numerical and graphical results imply that the proposed hybrid FA-GA is potentially more powerful when applied for the optimization problem considered here.

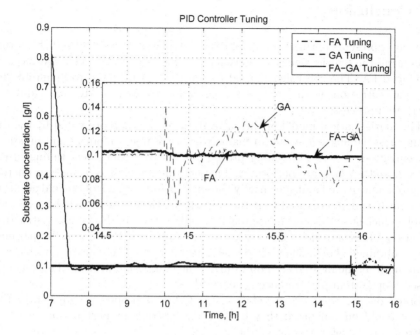

Fig. 3. Control variable - glucose concentration

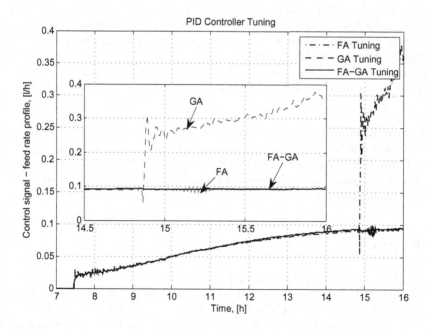

Fig. 4. Control signal - feed rate profiles

5   Conclusion

The paper presents optimal tuning of PID controller with Smith Predictor structure, using the hybrid algorithm between two meta-heuristics: GA and recently developed FA. The controller is used to control feed rate and to maintain glucose concentration at the desired set point for an *E. coli* MC4110 fed-batch cultivation process.

The mathematical model of the cultivation process is represented by the dynamic mass balance equations for main process variables - biomass and substrate concentration. A series of tuning procedures for PID controllers tuning, using FA, GA and FA-GA, are performed. The meta-heuristic algorithms' parameters are problem-oriented and specifically chosen to achieve an adequate and accurate decision. It is demonstrated that the meta-heuristics - pure and hybrid provide simple, efficient and accurate approach of tuning the Smith Predictor structure based on PID controller. As a result, a set of optimal PID controller parameters is obtained. For a short time, the controllers set the control variable and maintain it at the desired set-point during the cultivation process. Thus, a good closed-loop system performance is achieved.

Based on the comparison between FA, GA and proposed here hybrid FA-GA, it could be concluded that FA-GA, shows superior performance for PID controller parameter tuning of the considered nonlinear control system. The results show that the FA-GA takes the advantages of both FA's and GA's search ability, hence enhances the overall search ability and computational efficiency. The hybrid algorithm achieves the less value for the integrated square error criterion used here. Moreover the results are obtained for about two times less number of function evaluations and for about 27% less computational time.

Finally, it is shown that the PID controller tuning using FA-GA can be considered as an effective approach for the achievement of high quality and better performance of the designed control system for cultivation processes.

Acknowledgments. This work has been partially supported by the Bulgarian National Science Fund under the Grants DID 02/29 "Modelling Processes with Fixed Development Rules (ModProFix)" and DMU 02/4 "High quality control of biotechnological processes with application of modified conventional and metaheuristics methods".

References

1. Akpinar, S., Bayhan, G.M.: A Hybrid Genetic Aalgorithm for Mixed Model Assembly Line Balancing Problem with Parallel Workstations and Zoning Constraints. Engineering Applications of Artificial Intelligence 24(3), 449–457 (2011)
2. Al-Duwaish, H.N.: A Genetic Approach to the Identification of Linear Dynamical Systems with Static Nonlinearities. International Journal of Systems Science 31(3), 307–313 (2000)

3. Arndt, M., Hitzmann, B.: Feed Forward/feedback Control of Glucose Concentration during Cultivation of Escherichia coli. In: 8th IFAC Int. Conf. on Comp. Appl. in Biotechn, pp. 425–429 (2001)
4. Aström, K., Hagglund, T.: Advanced PID Control. Instrument Society of America (2006)
5. Benjamin, K.K., Ammanuel, A.N., David, A., Benjamin, Y.K.: Genetic Algorithm using for a Batch Fermentation Process Identification. Journal of Applied Sciences 8(12), 2272–2278 (2008)
6. Bonabeau, E., Dorigo, M., Theraulaz, G.: Swarm Intelligence: From Natural to Artificial Systems. Oxford University Press, New York (1999)
7. Brownlee, J.: Clever Algorithms. Nature-Inspired Programming Recipes, LuLu (2011)
8. da Silva, M.F.J., Perez, J.M.S., Pulido, J.A.G., Rodriguez, M.A.V.: AlineaGA - A Genetic Algorithm with Local Search Optimization for Multiple Sequence Alignment. Appl. Intell. 32, 164–172 (2010)
9. Galvez-Carrillo, M., De Keyser, R., Ionescu, C.: Application of a Smith Predictor based Nonlinear Predictive Controller to a Solar Power Plant. In: 7th IFAC Symposium on Nonlinear Control Systems, Pretoria, South Africa, August 21-24, pp. 188–193 (2007)
10. Garipov, E.: PID Controllers. Automatics and Informatics, vol. 3 (2006) (in Bulgarian)
11. Goldberg, D.E.: Genetic Algorithms in Search, Optimization and Machine Learning. Addison Wesley Longman, London (2006)
12. Gundogdu, O.: Optimal-tuning of PID Controller Gains using Genetic Algorithms. Journal of Engineering Sciences 11(1), 131–135 (2005)
13. Holland, J.H.: Adaptation in Natural and Artificial Systems, 2nd edn. MIT Press, Cambridge (1992)
14. Kim, J.S., Kim, J.-H., Park, J.-M., Park, S.-M., Choe, W.-Y., Heo, H.: Auto Tuning PID Controller based on Improved Genetic Algorithm for Reverse Osmosis Plant. World Academy of Science, Engineering and Technology 47, 384–389 (2008)
15. Kumar, S.M.G., Rakesh, B., Anantharaman, N.: Design of Controller using Simulated Annealing for a Real Time Process. International Journal of Computer Applications 2, 1053–1368 (2010)
16. Paplinski, J.P.: The Genetic Algorithm with Simplex Crossover for Identification of Time Delays. Intelligent Information Systems, pp. 337–346 (2010)
17. Puangdownreong, D., Kulworawanichpong, T., Sujitjorn, S.: Input weighting optimization for PID controllers based on the adaptive tabu search. IEEE TENCON 4, 451–454 (2004)
18. Ranganath, M., Renganathan, S., Gokulnath, C.: Identification of Bioprocesses using Genetic Algorithm. Bioprocess Engineering 21, 123–127 (1999)
19. Roeva, O.: Improvement of Genetic Algorithm Performance for Identification of Cultivation Process Models. In: Advances Topics on Evolutionary Computing, Book Series: Artificial Intelligence Series-WSEAS, pp. 34–39 (2008)
20. Roeva, O., Slavov, T.: Firefly Algorithm Tuning of PID Controller for Glucose Concentration Control during E. coli Fed-batch Cultivation Process. In: Federated Conference on Computer Science and Information Systems, WCO 2012, Poland, pp. 455–462 (2012)
21. Roeva, O., Slavov, T.: Fed-Batch Cultivation Control Based on Genetic Algorithm PID Controller Tuning. In: Dimov, I., Dimova, S., Kolkovska, N. (eds.) NMA 2010. LNCS, vol. 6046, pp. 289–296. Springer, Heidelberg (2011)

22. Slavov, T., Roeva, O.: Genetic Algorithm Tuning of PID Controller in Smith Predictor for Glucose Concentration Control. Int. J. Bioautomation 15(2), 101–114 (2011)
23. Smith, O.J.M.: A Controller to Overcome Dead Time. ISA Journal 6, 28–33 (1959)
24. Syam, W.P., Al-Harkan, I.M.: Comparison of Three Meta Heuristics to Optimize Hybrid Flow Shop Scheduling Problem with Parallel Machines. World Academy of Science, Engineering and Technology 62, 271–278 (2010)
25. Tahouni, N., Smith, R., Panjeshahi, M.H.: Comparison of Stochastic Methods with Respect to Performance and Reliability of Low-temperature Gas Separation Processes. The Canadian Journal of Chemical Engineering 88(2), 256–267 (2010)
26. Tseng, L.-Y., Lin, Y.-T.: A Hybrid Genetic Local Search Algorithm for the Permutation Flowshop Scheduling Problem. Europen J. of Operational Res. 198(1), 84–92 (2009)
27. Wang, Q., Spronck, P., Tracht, R.: An Overview of Genetic Algorithms Applied to Control Engineering Problems. Machine Learning and Cybernetics 3, 1651–1656 (2003)
28. Yang, X.S.: Nature-inspired Meta-heuristic Algorithms. Luniver Press, Beckington (2008)
29. Yang, X.-S.: Firefly Algorithms for Multimodal Optimization. In: Watanabe, O., Zeugmann, T. (eds.) SAGA 2009. LNCS, vol. 5792, pp. 169–178. Springer, Heidelberg (2009)
30. Yang, X.S.: Firefly Algorithm, Stochastic Test Functions and Design Optimisation. International Journal of Bio-inspired Computation 2(2), 78–84 (2010)

Some Properties of the Broyden Restricted Class of Updates with Oblique Projections

Andrzej Stachurski

Institute of Control and Computation Engineering
Warsaw University of Technology
Nowowiejska 15/19, 00-665 Warsaw, Poland
A.Stachurski@ia.pw.edu.pl

Abstract. In the paper the new formulation of the Broyden restricted convex class of updates involving oblique projections and some of its properties are presented. The new formulation involves two oblique projections. The new formula is a sum of two terms. The first one have the product form similar to that known for years for the famous BFGS (Broyden, Fletcher, Goldfarb, Shanno) update. The difference is that the oblique projection in the product contains vector defined as the convex, linear combination of the difference between consecutive iterative points and the image of the previous inverse hessian approximation on the corresponding difference of derivatives, i.e. gradients. The second standard term ensuring verification of the quasi-Newton condition is also an oblique projection multiplied by appropriate scalar. The formula relating the scalar parameter in the presented new version of updates with the formula appearing in the standard formulation is introduced and analyzed analytically and graphically. Formal proof of the theoretical equivalence of both updating formulas, when this relation is verified, is presented.

Some preliminary numerical experiments results on two twice continuously differentiable, strictly convex functions with increasing dimension are included.

1 Introduction

Problem considered in the current paper is the unconstrained minimization of a sufficiently smooth function

$$\min_{\mathbf{x} \in \mathcal{R}^n} f(\mathbf{x}) \tag{1}$$

Solution method assumes that given starting point $\mathbf{x}^0$ every consecutive approximate solution point is generated according to the following iterative formula

$$\mathbf{x}^{k+1} = \mathbf{x}^k + \alpha^k \mathbf{d}^k, \quad \forall k \geq 0 \tag{2}$$

where $\alpha^k > 0$ is the stepsize coefficient found in the directional minimization and the search direction $\mathbf{d}^k$ is equal to

$$\mathbf{d}^k = -\mathbf{H}^k \mathbf{g}^k \tag{3}$$

S. Fidanova (Ed.): *Recent Advances in Computational Optimization*, SCI 470, pp. 169–182.
DOI: 10.1007/978-3-319-00410-5_10   © Springer International Publishing Switzerland 2013

Matrix $\mathbf{H}^{k+1}$ is calculated at each step with the aid of vectors: $\mathbf{s}^k = \mathbf{x}^{k+1} - \mathbf{x}^k$, $\mathbf{r}^k = \nabla f(\mathbf{x}^{k+1}) - \nabla f(\mathbf{x}^k)$ and the previous matrix $\mathbf{H}^k$.

Problems of unconstrained functions minimization arise first of all as the result of the least squares approach to solve sets of nonlinear equations (see for instance the problem of determining stresses in RC ring sections with openings in Lechman and Stachurski [11] and Stachurski and Lechman [20]) and identification of parameters appearing in the model in a nonlinear way (as for instance in the augmented Gurson model describing the creation and growth of voids in the porous material considered in Nowak and Stachurski in the sequence of publications [12]- [16]).

Broyden convex class of updates is usually expressed in the following way (see for instance Sun and Yuan [25])

$$\mathbf{H}^{k+1} = \mathbf{H}^k + \left(1 + \Phi \frac{\left(\mathbf{r}^k\right)^T \mathbf{H}^k \mathbf{r}^k}{\left(\mathbf{r}^k\right)^T \mathbf{s}^k}\right) \frac{\mathbf{s}^k \left(\mathbf{s}^k\right)^T}{\left(\mathbf{r}^k\right)^T \mathbf{s}^k}$$
$$- (1 - \Phi) \frac{\mathbf{H}^k \mathbf{r}^k \left(\mathbf{r}^k\right)^T \mathbf{H}^k}{\left(\mathbf{r}^k\right)^T \mathbf{H}^k \mathbf{r}^k} \tag{4}$$
$$- \Phi \frac{\mathbf{s}^k \left(\mathbf{r}^k\right)^T \mathbf{H}^k + \mathbf{H}^k \mathbf{r}^k \left(\mathbf{s}^k\right)^T}{\left(\mathbf{r}^k\right)^T \mathbf{s}^k}$$

where Φ is a scalar belonging to the interval $[0, 1]$.

In the consecutive section 2 we shall show an alternate updating formula of the form

$$\mathbf{H}^{k+1} = \mathbf{P}^T \mathbf{H}^k \mathbf{P} + \beta \mathbf{Q}$$

where $\mathbf{P}$ and $\mathbf{Q}$ are oblique projections, i.e. $\mathbf{P}$ sets to null any vector collinear with $\mathbf{r}^k$ and $\mathbf{Q}$ nullifies any vector orthogonal to $\mathbf{s}^k$ and $\mathbf{P}\mathbf{P} = \mathbf{P}$ and $\mathbf{Q}\mathbf{Q} = \mathbf{Q}$. Parameter β is a positive scalar changing from one iteration to another. Reader interested in the theory of oblique projections and their properties may find more information for instance in Afriat [1] or Szyld [24]. In section 3 we show formula relating scalar parameters in both formulas. We present formal theorem with the proof of the equivalence of the new form of updates and the standard formulation when the aforementioned relation holds. In consecutive section 4 we analyze the relating formula analytically and graphically. The behavior depends heavily on the ratio $\frac{\mathbf{r}_k^T \mathbf{H}^k \mathbf{s}_k}{\mathbf{r}_k^T \mathbf{s}_k}$.

Section 5 contains some preliminary computational results obtained with the aid of quasi-newton methods with updates defined by the discussed formula with parameter $\Theta = 1$, 0 and $\frac{1}{2}$. Testing examples are two strictly convex functions constructed in the way permitting easily increase their dimensions. In the last section 6 conclusions and comments following from the numerical experiments and of general theoretical character are presented.

2   Oblique Projections in the Formula of the Broyden Class Updates

Broyden convex class may be equivalently represented by the following updating formula

$$\mathbf{H}^{k+1} = \left(\mathbf{P}^k\right)^T \mathbf{H}^k \mathbf{P}^k + \frac{\mathbf{s}^k \left(\mathbf{s}^k\right)^T}{\left(\mathbf{r}^k\right)^T \mathbf{s}^k} \tag{5}$$

where $\mathbf{P}^k$ is the projection matrix defined as follows

$$\mathbf{P}^k = \mathbf{I} - \frac{\mathbf{r}^k \left[\Theta \mathbf{s}^k + (1-\Theta)\mathbf{H}^k \mathbf{r}^k\right]^T}{\left(\mathbf{r}^k\right)^T \left(\Theta \mathbf{s}^k + (1-\Theta)\mathbf{H}^k \mathbf{r}^k\right)} \tag{6}$$

and parameter $\Theta \in [0,1]$.

2.1   Involved Oblique Projections

First, let's show that $\mathbf{P}^k$ is the projection matrix transforming vector $\mathbf{r}^k$ to the null vector $\mathbf{0}$

$$\begin{aligned}
\mathbf{P}^k \mathbf{r}^k &= \left(\mathbf{I} - \frac{\mathbf{r}^k \left[\Theta \mathbf{s}^k + (1-\Theta)\mathbf{H}^k \mathbf{r}^k\right]^T}{\left(\mathbf{r}^k\right)^T \left(\Theta \mathbf{s}^k + (1-\Theta)\mathbf{H}^k \mathbf{r}^k\right)}\right) \mathbf{r}^k \\
&= \mathbf{r}^k - \mathbf{r}^k \frac{\left[\Theta \mathbf{s}^k + (1-\Theta)\mathbf{H}^k \mathbf{r}^k\right]^T \mathbf{r}^k}{\left(\mathbf{r}^k\right)^T \left(\Theta \mathbf{s}^k + (1-\Theta)\mathbf{H}^k \mathbf{r}^k\right)} = \mathbf{0}
\end{aligned} \tag{7}$$

Matrix $\mathbf{P}^k$ is an oblique projection (definition and properties of such projections may be found for instance in Afriat [1] or Szyld [24]), because

$$\begin{aligned}
\mathbf{P}^k \mathbf{P}^k &= \left(\mathbf{I} - \frac{\mathbf{r}^k \left[\Theta \mathbf{s}^k + (1-\Theta)\mathbf{H}^k \mathbf{r}^k\right]^T}{\left(\mathbf{r}^k\right)^T \left(\Theta \mathbf{s}^k + (1-\Theta)\mathbf{H}^k \mathbf{r}^k\right)}\right) \left(\mathbf{I} - \frac{\mathbf{r}^k \left[\Theta \mathbf{s}^k + (1-\Theta)\mathbf{H}^k \mathbf{r}^k\right]^T}{\left(\mathbf{r}^k\right)^T \left(\Theta \mathbf{s}^k + (1-\Theta)\mathbf{H}^k \mathbf{r}^k\right)}\right) \\
&= \mathbf{I} - \frac{\mathbf{r}^k \left[\Theta \mathbf{s}^k + (1-\Theta)\mathbf{H}^k \mathbf{r}^k\right]^T}{\left(\mathbf{r}^k\right)^T \left(\Theta \mathbf{s}^k + (1-\Theta)\mathbf{H}^k \mathbf{r}^k\right)} \\
&\quad + \frac{\mathbf{r}^k \left[\Theta \mathbf{s}^k + (1-\Theta)\mathbf{H}^k \mathbf{r}^k\right]^T \left(\mathbf{r}^k\right)^T \left(\Theta \mathbf{s}^k + (1-\Theta)\mathbf{H}^k \mathbf{r}^k\right)}{\left(\mathbf{r}^k\right)^T \left(\Theta \mathbf{s}^k + (1-\Theta)\mathbf{H}^k \mathbf{r}^k\right) \left(\mathbf{r}^k\right)^T \left(\Theta \mathbf{s}^k + (1-\Theta)\mathbf{H}^k \mathbf{r}^k\right)} \\
&\quad - \frac{\mathbf{r}^k \left[\Theta \mathbf{s}^k + (1-\Theta)\mathbf{H}^k \mathbf{r}^k\right]^T}{\left(\mathbf{r}^k\right)^T \left(\Theta \mathbf{s}^k + (1-\Theta)\mathbf{H}^k \mathbf{r}^k\right)} \\
&= \mathbf{I} - \frac{\mathbf{r}^k \left[\Theta \mathbf{s}^k + (1-\Theta)\mathbf{H}^k \mathbf{r}^k\right]^T}{\left(\mathbf{r}^k\right)^T \left(\Theta \mathbf{s}^k + (1-\Theta)\mathbf{H}^k \mathbf{r}^k\right)} = \mathbf{P}^k
\end{aligned}$$

Second term in formula (5) is also an oblique projection

$$\frac{\mathbf{s}^k \left(\mathbf{s}^k\right)^T}{\|\mathbf{s}^k\|^2} \tag{8}$$

multiplied by a scalar

$$\beta = \frac{\|\mathbf{s}^k\|^2}{\left(\mathbf{r}^k\right)^T \mathbf{s}^k} \tag{9}$$

It is not difficult to show that formula (8) defines an oblique projection.

Representation (5) has appeared for the first time in Stachurski [22]. It was proposed there as a new quasi-newton update. Later the author has realized that it's a new representation of the famous convex class of Broyden proposed for the first time in [3].

2.2   BFGS Update and Oblique Projections

Similar product representation is known for many years for the BFGS update (name derived from the family names of its authors Broyden [3], Fletcher [8], Goldfarb [10] and Shanno [17])

$$\begin{aligned}
\mathbf{H}^{k+1}_{BFGS} = \mathbf{H}^k &+ \left(1 + \frac{\left(\mathbf{r}^k\right)^T \mathbf{H}^k \mathbf{r}^k}{\left(\mathbf{r}^k\right)^T \mathbf{s}^k}\right) \frac{\mathbf{s}^k \left(\mathbf{s}^k\right)^T}{\left(\mathbf{r}^k\right)^T \mathbf{s}^k} \\
&- \frac{\mathbf{s}^k \left(\mathbf{r}^k\right)^T \mathbf{H}^k + \mathbf{H}^k \mathbf{r}^k \left(\mathbf{s}^k\right)^T}{\left(\mathbf{r}^k\right)^T \mathbf{s}^k}
\end{aligned} \tag{10}$$

In the limited memory BFGS method (see for instance Xiao [26]) the following representation of the BFGS update is frequently used

$$\mathbf{H}^{k+1} = \left(\mathbf{I} - \frac{\mathbf{r}^k \left(\mathbf{s}^k\right)^T}{\left(\mathbf{s}^k\right)^T \mathbf{r}^k}\right)^T \mathbf{H}^k \left(\mathbf{I} - \frac{\mathbf{r}^k \left(\mathbf{s}^k\right)^T}{\left(\mathbf{s}^k\right)^T \mathbf{r}^k}\right) + \frac{\mathbf{s}^k \left(\mathbf{s}^k\right)^T}{\left(\mathbf{r}^k\right)^T \mathbf{s}^k} \tag{11}$$

It is easy to notice that formula (11) is represented by formulae (5) and (6) with $\Theta = 1$.

2.3   DFP Update and Oblique Projections

The second famous update – DFP (proposed originally by Davidon [5] and [6] and further developed by Fletcher and Powell [7])

$$\mathbf{H}^{k+1}_{DFP} = \mathbf{H}^k - \frac{\mathbf{H}^k \mathbf{r}^k \left(\mathbf{r}^k\right)^T \mathbf{H}^k}{\left(\mathbf{r}^k\right)^T \mathbf{H}^k \mathbf{r}^k} + \frac{\mathbf{s}^k \left(\mathbf{s}^k\right)^T}{\left(\mathbf{r}^k\right)^T \mathbf{s}^k} \tag{12}$$

may be also expressed with the aid of oblique projections as follows

$$\mathbf{H}^{k+1} = \left(\mathbf{I} - \frac{\mathbf{r}^k \left(\mathbf{H}^k \mathbf{r}^k\right)^T}{\left(\mathbf{H}^k \mathbf{r}^k\right)^T \mathbf{r}^k}\right)^T \mathbf{H}^k \left(\mathbf{I} - \frac{\mathbf{r}^k \left(\mathbf{H}^k \mathbf{r}^k\right)^T}{\left(\mathbf{H}^k \mathbf{r}^k\right)^T \mathbf{r}^k}\right) + \frac{\mathbf{s}^k \left(\mathbf{s}^k\right)^T}{\left(\mathbf{r}^k\right)^T \mathbf{s}^k} \tag{13}$$

It is easy to observe that formula (11) is represented by formulas (5) and (6) with $\Theta = 0$. Equivalence of formulas (12) and (13) was shown for the first time in Stachurski [21].

3   Relation between the Scalar Parameters of Updates

Parameters Φ and Θ are mutually connected by the following formula

$$\Phi = \Theta^2 \frac{\left(\mathbf{r}_k^T \mathbf{s}_k\right)^2}{\left(\mathbf{r}_k^T \mathbf{u}_k\right)^2} \tag{14}$$

where vector $\mathbf{u}_k = \Theta \mathbf{s}_k + (1 - \Theta)\mathbf{H}^k \mathbf{r}_k$.

Formal prove showing that when equality (14) holds then formulas (4) and (5) are equivalent is presented below.

Theorem 1. *Let parameters* $\Theta \in [0,1]$ *and* $\Phi \in [0,1]$ *verify equality (14) and* $\mathbf{r}_k^T \mathbf{s}_k > 0$.

Then in the exact arithmetic the family of updates with oblique projections defined by formulae (5–6) and the Broyden's restricted class (4) produce identical matrices $\mathbf{H}^{k+1}$.

Proof. Let's rewrite the updating formula with oblique projections (5-6) in the following equivalent way

$$\begin{aligned}
\mathbf{H}^{k+1} = \mathbf{H}^k &+ \left[-\Theta + \Theta(1-\Theta)\frac{\mathbf{r}_k^T \mathbf{H}^k \mathbf{r}_k}{\mathbf{r}_k^T \mathbf{u}_k}\right]\frac{\mathbf{H}^k \mathbf{r}_k \mathbf{s}_k^T + \mathbf{s}_k \mathbf{r}_k^T \mathbf{H}^k}{\mathbf{r}_k^T \mathbf{u}_k} \\
&+ \left[(1-\Theta)^2 \frac{\mathbf{r}_k^T \mathbf{H}^k \mathbf{r}_k}{\mathbf{r}_k^T \mathbf{u}_k} - 2(1-\Theta)\right]\frac{\mathbf{H}^k \mathbf{r}_k \mathbf{r}_k^T \mathbf{H}^k}{\mathbf{r}_k^T \mathbf{u}_k} \\
&+ \left[\frac{\Theta^2}{\mathbf{r}_k^T \mathbf{u}_k}\frac{\mathbf{r}_k^T \mathbf{H}^k \mathbf{r}_k}{\mathbf{r}_k^T \mathbf{u}_k} + \frac{1}{\mathbf{r}_k^T \mathbf{s}_k}\right]\mathbf{s}_k \mathbf{s}_k^T
\end{aligned} \tag{15}$$

where

$$\mathbf{u}_k = \Theta \mathbf{s}_k + (1-\Theta)\mathbf{H}^k \mathbf{r}_k$$

Similarly, the Broyden's restricted class updating formula can be expressed equivalently as follows

$$\begin{aligned}
\mathbf{H}^{k+1} = \mathbf{H}^k &- \Phi \frac{\mathbf{s}_k \mathbf{r}_k^T \mathbf{H}^k + \mathbf{H}^k \mathbf{r}_k \mathbf{s}_k^T}{\mathbf{r}_k^T \mathbf{s}_k} \\
&- (1-\Phi)\frac{\mathbf{H}^k \mathbf{r}_k \mathbf{r}_k^T \mathbf{H}^k}{\mathbf{r}_k^T \mathbf{H}^k \mathbf{r}_k} \\
&+ \left(\Phi \frac{\mathbf{r}_k^T \mathbf{H}^k \mathbf{r}_k}{\left(\mathbf{r}_k^T \mathbf{s}_k\right)^2} + \frac{1}{\mathbf{r}_k^T \mathbf{s}_k}\right)\mathbf{s}_k \mathbf{s}_k^T
\end{aligned} \tag{16}$$

Now, we shall show that the coefficients in (15) and (16) associated with the terms $\mathbf{H}^k \mathbf{r}_k \mathbf{s}_k^T + \mathbf{H}^k \mathbf{r}_k \mathbf{s}_k^T$, $\mathbf{H}^k \mathbf{r}_k \mathbf{r}_k^T \mathbf{H}^k$ and $\mathbf{s}_k \mathbf{s}_k^T$ are equal when formula (14) is verified, i.e. we shall prove validity of the following three equalities

$$\begin{aligned}
-\frac{\Theta}{\mathbf{r}_k^T \mathbf{u}_k} + \Theta(1-\Theta)\frac{\mathbf{r}_k^T \mathbf{H}^k \mathbf{r}_k}{\left(\mathbf{r}_k^T \mathbf{u}_k\right)^2} &= -\frac{\Phi}{\mathbf{r}_k^T \mathbf{s}_k} \\
(1-\Theta)^2 \frac{\mathbf{r}_k^T \mathbf{H}^k \mathbf{r}_k}{\left(\mathbf{r}_k^T \mathbf{u}_k\right)^2} - \frac{2(1-\Theta)}{\mathbf{r}_k^T \mathbf{u}_k} &= -\frac{(1-\Phi)}{\mathbf{r}_k^T \mathbf{H}^k \mathbf{r}_k} \\
\frac{\Theta^2}{\mathbf{r}_k^T \mathbf{u}_k}\frac{\mathbf{r}_k^T \mathbf{H}^k \mathbf{r}_k}{\mathbf{r}_k^T \mathbf{u}_k} + \frac{1}{\mathbf{r}_k^T \mathbf{s}_k} &= \frac{1}{\mathbf{r}_k^T \mathbf{s}_k} + \Phi \frac{\mathbf{r}_k^T \mathbf{H}^k \mathbf{r}_k}{\left(\mathbf{r}_k^T \mathbf{s}_k\right)^2}
\end{aligned} \tag{17}$$

Let's denote the left-hand sides coefficients in equation (17), in the first line by N_1, second by N_2 and the third by N_3.

It is readily seen that coefficients in the third lines are equal

$$\frac{\Theta^2}{\mathbf{r}_k^T\mathbf{u}_k}\frac{\mathbf{r}_k^T\mathbf{H}^k\mathbf{r}_k}{\mathbf{r}_k^T\mathbf{u}_k} + \frac{1}{\mathbf{r}_k^T\mathbf{s}_k} = \Phi\frac{\mathbf{r}_k^T\mathbf{H}^k\mathbf{r}_k}{\left(\mathbf{r}_k^T\mathbf{s}_k\right)^2} + \frac{1}{\mathbf{r}_k^T\mathbf{s}_k}$$

It suffices to subtract $\frac{1}{\mathbf{r}_k^T\mathbf{s}_k}$ from both sides, multiply by $\left(\mathbf{r}_k^T\mathbf{s}_k\right)^2$ and divide by $\mathbf{r}_k^T\mathbf{H}^k\mathbf{r}_k$.

Let us now consider the N_1 coefficient, appearing in the first line of formula (15)

$$N_1 = -\frac{\Theta}{\mathbf{r}_k^T\mathbf{u}_k} + \Theta(1-\Theta)\frac{\mathbf{r}_k^T\mathbf{H}^k\mathbf{r}_k}{\left(\mathbf{r}_k^T\mathbf{u}_k\right)^2}$$

Transformation of the fractions to the form with the common denominator and their addition leads to the following formulation

$$N_1 = \frac{-\Theta\mathbf{r}_k^T\mathbf{u}_k + \Theta(1-\Theta)\mathbf{r}_k^T\mathbf{H}^k\mathbf{r}_k}{\left(\mathbf{r}_k^T\mathbf{u}_k\right)^2}$$

Applying the definition of vector $\mathbf{u}_k = \Theta\mathbf{s}_k + (1-\Theta)\mathbf{H}^k\mathbf{r}_k$ we obtain

$$N_1 = \frac{-\Theta^2\mathbf{r}_k^T\mathbf{s}_k - \Theta(1-\Theta)\mathbf{r}_k^T\mathbf{H}^k\mathbf{r}_k + \Theta(1-\Theta)\mathbf{r}_k^T\mathbf{H}^k\mathbf{r}_k}{\left(\mathbf{r}_k^T\mathbf{u}_k\right)^2}$$

Now, reduction of similar terms in the numerator yields

$$N_1 = \frac{-\Theta^2\mathbf{r}_k^T\mathbf{s}_k}{\left(\mathbf{r}_k^T\mathbf{u}_k\right)^2} = -\frac{\Phi}{\mathbf{r}_k^T\mathbf{s}_k}$$

Let's analyze the N_2 coefficient

$$N_2 = (1-\Theta)^2\frac{\mathbf{r}_k^T\mathbf{H}^k\mathbf{r}_k}{\left(\mathbf{r}_k^T\mathbf{u}_k\right)^2} - \frac{2(1-\Theta)}{\mathbf{r}_k^T\mathbf{u}_k}$$

First, let's convert the two fractions to the form with the common denominator, subtract them and multiply the resulting numerator and denominator by the same factor $\mathbf{r}_k^T\mathbf{H}^k\mathbf{r}_k$. After doing that we obtain the following representation

$$N_2 = \frac{(1-\Theta)^2\left(\mathbf{r}_k^T\mathbf{H}^k\mathbf{r}_k\right)^2 - 2(1-\Theta)\mathbf{r}_k^T\mathbf{H}^k\mathbf{r}_k\mathbf{r}_k^T\mathbf{u}_k}{\mathbf{r}_k^T\mathbf{H}^k\mathbf{r}_k\left(\mathbf{r}_k^T\mathbf{u}_k\right)^2}$$

Now, application of the identity

$$\begin{aligned}\left(\Theta\mathbf{r}_k^T\mathbf{s}_k\right)^2 &= \left(\mathbf{r}_k^T\mathbf{u}_k - (1-\Theta)\mathbf{r}_k^T\mathbf{H}^k\mathbf{r}_k\right)^2 \\ &= \left(\mathbf{r}_k^T\mathbf{u}_k\right)^2 - 2(1-\Theta)\mathbf{r}_k^T\mathbf{H}^k\mathbf{r}_k\mathbf{r}_k^T\mathbf{u}_k + (1-\Theta)^2\left(\mathbf{r}_k^T\mathbf{H}^k\mathbf{r}_k\right)^2\end{aligned}$$

in the numerator yields

$$N_2 = \frac{\Theta^2 \left(\mathbf{r}_k^T \mathbf{s}_k\right)^2 - \left(\mathbf{r}_k^T \mathbf{u}_k\right)^2}{\mathbf{r}_k^T \mathbf{H}^k \mathbf{r}_k \left(\mathbf{r}_k^T \mathbf{u}_k\right)^2}$$

Division of the numerator and denominator by quantity $\left(\mathbf{r}_k^T \mathbf{u}_k\right)^2$ produces the result

$$N_2 = \frac{\Theta^2 \left(\dfrac{\mathbf{r}_k^T \mathbf{s}_k}{\mathbf{r}_k^T \mathbf{u}_k}\right)^2 - 1}{\mathbf{r}_k^T \mathbf{H}^k \mathbf{r}_k}$$

Taking into account the assumption we obtain that

$$N_2 = -\frac{1 - \Phi}{\mathbf{r}_k^T \mathbf{H}^k \mathbf{r}_k}$$

what concludes the proof.

4 Properties of the Relation between Parameters

Formula (14) relating parameters Φ and Θ may be rewritten as follows

$$\Phi = \Theta^2 \left[\frac{\mathbf{r}_k^T \mathbf{s}_k}{\Theta \mathbf{r}_k^T \mathbf{H}^k \mathbf{r}_k + (1 - \Theta)\mathbf{r}_k^T \mathbf{s}_k}\right]^2 \tag{18}$$

Division of numerator and denominator of the second fraction by $\mathbf{r}_k^T \mathbf{s}^k$ yields

$$\Phi = \Theta^2 \left[\frac{1}{\Theta + (1 - \Theta)\dfrac{\mathbf{r}_k^T \mathbf{H}^k \mathbf{r}_k}{\mathbf{r}_k^T \mathbf{s}_k}}\right]^2 \tag{19}$$

Now let's denote by a the ratio present in the denominator

$$a = \frac{\mathbf{r}_k^T \mathbf{H}^k \mathbf{r}_k}{\mathbf{r}_k^T \mathbf{s}_k}$$

what leads to the following dependence

$$\Phi(\Theta) = \Theta^2 \left[\frac{1}{\Theta + (1 - \Theta)a}\right]^2 \tag{20}$$

Let's fix some $\Theta \in (0, 1)$ and calculated the limits for parameter $a \longrightarrow 0$

$$\lim_{a \longrightarrow 0} \Phi(\Theta) = \lim_{a \longrightarrow 0} \left[\frac{\Theta}{\Theta + (1 - \Theta)a}\right]^2 = 1 \tag{21}$$

and for $a \longrightarrow \infty$

$$\lim_{a \longrightarrow \infty} \Phi(\Theta) = \lim_{a \longrightarrow \infty} \left[\frac{\Theta}{\Theta + (1 - \Theta)a} \right]^2 = \lim_{a \longrightarrow \infty} \left[\frac{\Theta/a}{\Theta/a + 1 - \Theta} \right]^2 = 0 \quad (22)$$

Hence we may conclude that for small values of the ratio $\frac{\mathbf{r}_k^T \mathbf{H}^k \mathbf{r}_k}{\mathbf{r}_k^T \mathbf{s}_k}$ the new updating formula automatically approaches the BFGS formula. For large values of the the formula approaches the DFP update.

To realize what values of the ratio are significant we will analyze the dependence (20) graphically. Below we draw Φ as a function of Θ for different values of parameter a. The corresponding plots are presented on Fig. 1.

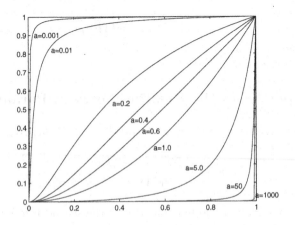

Fig. 1. Dependence of $\Phi(\Theta)$ for different values of the ratio a

Deeper look into the plots shows that with small values of ratio a, parameter Φ is almost equal to 1 for $\Theta > 0.1$ when ratio a drops to the value equal to 0.001. Large values of a results in Φ almost equal to 0 for $\Theta < 0.9$, when a reaches the level 100. For $\Theta = 1/2$ those levels are much closer: parameter Φ is close to 1 already for $a = 0.01$ and to 0 for $\Theta = 5$. This observation points that the updating formula has a built–in mechanism of selection of formulas close to DFP update, when $\mathbf{r}_k^T \mathbf{H}^k \mathbf{r}_k$ is significantly larger than $\mathbf{r}_k^T \mathbf{s}^k$ and BFGS update when it is significantly smaller. It is consistent with the Fletcher [8] switching rule between DFP and BFGS updates: select DFP update, when $\mathbf{r}_k^T \mathbf{H}^k \mathbf{r}_k > \mathbf{r}_k^T \mathbf{s}^k$ and BFGS update in the opposite case.

5   Numerical Experiments

In the current section the results of numerical calculations are presented. They are realized by means of three variants of updates specified by formulas (5-6) with parameter Θ equal to 1 (corresponding to the BFGS method), 0 (DFP method) and $\frac{1}{2}$. Three variants of directional minimization were tested: Armijo directional minimization ensuring verification of the Armijo condition (identical with the first Godstein test) (it is denoted below in the results table by A.), directional minimization ensuring verification of the Wolfe conditions (denoted below by W.), directional minimization ensuring verification of both Goldstein conditions (denoted by G. respectively).

The first directional minimization was realized by setting the starting value of the directional step–size and its consecutive reduction by some constant coefficient belonging to the interval $(0, 1)$ until the first Goldstein condition is met. In the second directional minimization Wolfe conditions were used as the stopping criterion and in the third variant two Goldstein tests served as the stopping criterion for the directional minimization. In the second and third variant of the directional minimization consecutive approximations of the step–size length were generated as the minimum point of the parabola approximating function $\tilde{f}(\alpha) = f(\mathbf{x}^k + \alpha * \mathbf{d}^k)$.

5.1   Test Functions

Two strictly convex, n-dimensional functions with increasing dimension were used for testing. Dimensions were equal to $n = 2, 10, 50, 100, 500, 1000, 2000$. The first was obtained by raising up to the second power values of a strictly convex quadratic function with positive values (its minimal value was positive)

$$f_1(\mathbf{x}) = \frac{1}{2} \left[f_{qua}(\mathbf{x}) \right]^2, \tag{23}$$

where

$$f_{qua}(\mathbf{x}) = \frac{1}{2} (\mathbf{x} - \mathbf{e})^T \mathbf{Q} (\mathbf{x} - \mathbf{e}) + 1.0 \tag{24}$$

Vector $\mathbf{e}^T = [1, 1, \ldots, 1]$ in formula (24) consists of ones, second derivative matrix $\mathbf{Q}$ was generated randomly (some extra operations ensuring its positive definiteness were involved). First, quadratic matrix $\bar{\mathbf{Q}}$ of size $n \times n$ is created by invoking MATLAB function *rand*. Its elements are numbers belonging to the interval $[0, 1]$. Next, lower triangular matrix $\mathbf{L}$ is created on the basis of matrix $\bar{\mathbf{Q}}$. It has entries with 0 values on the main diagonal and its entries below the main diagonal were identical with that of $\bar{\mathbf{Q}}$ matrix, i.e.

$$\begin{aligned} L_{ij} &= \bar{Q}_{ij}, \quad \forall i < j, \\ L_{ij} &= 0, \quad \forall i \geq j \end{aligned}$$

Finally, matrix $\mathbf{Q}$ is defined in the consecutive step by the formula

$$\tilde{\mathbf{Q}} = \frac{1}{2} \left[\mathbf{L} + \mathbf{L}^T \right]$$

to which the diagonal matrix $\mathbf{D}$ defined as

$$D_{ii} = \sum_{j=1}^{n} \tilde{Q}_{ij} + 1, \quad i = 1, \ldots n$$

is added. The resulting matrix $\mathbf{Q} = \tilde{\mathbf{Q}} + \mathbf{D}$ was a diagonally dominated matrix with nonnegative entries. All entries outside the main diagonal belong to the interval $[0, 1]$. Furthermore, it was positive definite. The last property was checked numerically to verify correctness.

Vector $\mathbf{e}$ is the unique, local and global minimum of the function (23) constructed in this way. Its optimal value is $1/2$. Described construction makes use of the random numbers generator, however data defining the generated problem of a given size together with the starting point were stored in the MATLAB data file with extension *.dat* by means of the *save* command. The file has been loaded to the operational memory during the start of any method by means of the *load* command. It ensures compatibility of the computational results for various methods which were tested. For any assumed dimension – 2, 10, 50, 100, 500, 1000, 2000 an independent problem has been generated. In any case, the way of generating the problem was identical with that described above. Similarly, the starting points were the same for any method for the problem of the given dimension. All calculations were run on the 32-bit personal computer with processor *Intel(R) Pentium(R) 4 CPU 3.20GHz*, with RAM memory of 1GB capacity, working under the Windows XP Professional operating system.

In the second example function f_2 is generated similarly. The only difference is that instead of taking the second power of the quadratic function f_{qua} we assume its natural logarithm, i.e.

$$f_2(\mathbf{x}) = \ln(f_{qua}(\mathbf{x})) \tag{25}$$

5.2   Results Obtained by Means of Selected Members of the Broyden Convex Class

Three variants of Broyden methods belonging to the convex class in version with oblique projections defined by (5) were implemented. The selected three variants are: BFGS method with $\Theta = 1$, DFP method with $\Theta = 0$ and the third version with $\Theta = \frac{1}{2}$. Every method was implemented with three above mentioned directional minimizations: Armijo (A.), Wolfe (W.) and Goldstein (G.). Hence we considered altogether nine variants of methods. Every method variant has been run with the same MATLAB m-function implementing the directional minimization and on the same set of test problems with increasing dimension, generated as described above. Stopping criteria were also the same - on the derivative norm and on the minimized function value. Let's notice that we know the optimum goal function value. The results for goal function f_1 are collected in table 1 and for the second goal function f_2 in table 2.

Symbol (M) placed instead of the number of iterations means stop due to exceeding the maximal number of iterations, set by the user. Symbol (P) denotes the user break by pressing simultaneously combinations of keys CTRL-C and (O) stopping the calculations due to the zero value in the denominator in the updating formula (something theoretically impossible in the exact arithmetic, but on the computer we never carry out calculations in the exact arithmetic). Later appropriate safeguards were introduced.

Table 1. Number of iterations of Broyden convex class with different directional min-imization for function f_1

Dir.	Problem size $n =$						
min.	2	10	50	100	500	1000	2000
BFGS method ($\Theta = 1$)							
A.	5	(M)	(M)	(M)	(M)	(M)	–
W.	4	27	70	185	533	1080	821
G.	4	47	139	314	809	1728	1762
DFP method ($\Theta = 0$)							
A.	5	(M)	(M)	(M)	(M)	(M)	–
W.	4	29	76	313	5652	(M)	2986(O)
G.	4	68	249	(M)	(M)	(M)	(P)
variant with $\Theta = \frac{1}{2}$							
A.	5	(M)	(M)	(M)	(M)	(P)	–
W.	4	29	69	190	597	1443	1106
G.	4	58	185	582	1331	3254	3819

The obtained results prove the BFGS method superiority over the DFP and the third variant with $\Theta = \frac{1}{2}$. Furthermore, they have shown that for problems of larger dimension the Armijo directional minimization (i.e. decreasing the step–size from a given starting value by a constant coefficient until the first Goldstein test is met) is totally useless. The directional minimization with the Wolfe stopping conditions proved to be the best one. Number of iterations of the BFGS method implemented with the directional minimization ensuring verification of the Wolfe conditions was substantially smaller than in all other considered variants. The only exception were the problems of smallest size equal to 2.

Our calculations have shown a very important role of the round–off errors. Furthermore calculations were realized without any scaling. Hence the round–off errors may be the source of sometimes strange behaviour in some cases. For instance, for larger dimensions we observed sometimes the number of iterations smaller than the size of the problem. Let's stress that numbers smaller than 10^{-72} are treated as equal to 0 on the PC computer. Numbers are represented with a finite number of digits and number 10 in appropriate power. This may lead and led to some unexpected overflows. For instance, in theory, scalar product $r_k^T s_k > 0$ should be positive in our situation since the goal function is strictly convex and twice continuously differentiable. However, from time to time it was equal to 0 and it was necessary to introduce conditional updating and omit in such situation the update.

Table 2. Number of iterations of Broyden convex class with different directional minimization for function f_2

Dir.	Problem size $n =$						
min.	2	10	50	100	500	1000	2000
BFGS method ($\Theta = 1$)							
A.	23	51	314	627	2787	(M)	(M)
W.	4	17	37	48	116	406	768
G.	4	25	38	52	524	428	1655
DFP method ($\Theta = 0$)							
A.	23	51	311	(M)	(M)	(M)	–
W.	4	17	38	48	116	(M)	(M)
G.	4	20	53	67	289	419	1086
variant with $\Theta = \frac{1}{2}$							
A.	23	51	501	(M)	(M)	(M)	–
W.	4	17	40	48	116	(M)	(M)
G.	4	20	60	75	285	534	906

6   Conclusions and Comments

Updates representation with oblique projections gives a deeper look into the structure of the existing variable metric updates. It offers new possibilities in convergence analysis of quasi-newton methods for minimization. It would be then possible to exploit the existing rich algebraic theory of oblique projections. Furthermore it opens the possibility to exploit in context of the limited memory methods any member of the Broyden convex class. We are not restricted to the BFGS as it was up till now.

Analysis of formula (14) relating parameters Φ and Θ carried out in section 4 has shown that Φ is close to 1 for the majority of the Θ values lying in $(0, 1)$ interval when the ratio $a = \dfrac{\mathbf{r}_k^T \mathbf{H}^k \mathbf{r}_k}{\mathbf{r}_k^T \mathbf{s}_k}$ is small. Then the updating formula is close to the BFGS update. In the opposite situation when the ratio is large, then Φ approaches 0, which corresponds to the DFP update. Let's stress that it happens for relatively large ($a = 0.001$) in the first case and relatively small ($a = 50$) in the second.

We see as an interesting and open as yet problem of direct control of the smallest and largest eigenvalues of the updated matrix approximating the second order derivative or its inverse. This would simplify the existing convergence proofs for the quasi-Newton methods and help in obtaining some progress on their convergence for problems with hessian matrix of the goal function having singularities.

References

1. Afriat, S.N.: Orthogonal and oblique projectors and the characteristics of pairs of vector spaces. Proc. Camb. Philos. Soc. 53, 800–816 (1957)
2. Bazaraa, M.S., Sherali, J., Shetty, C.M.: Nonlinear Programming. Theory and Algorithms. John Wiley and Sons, New York (1993)
3. Broyden, C.G.: The convergence of a class double-rank minimization algorithms. Journal of the Institute of Mathematics and its Applications 6, 76–90 (1970)
4. Byrd, R.H., Nocedal, J., Yuan, Y.: Global Convergence of a Class of Variable Metric Algorithms. SIAM Journal on Numerical Analysis 24, 1171–1190 (1987)
5. Davidon, W.C.: Variable metric method for minimization. AEC Res. and Dev. Report, ANL-5990 (1959) (revised)
6. Davidon, W.C.: Variable metric method for minimization. SIAM J. on Optimization 1, 1–17 (1991)
7. Fletcher, R.: A rapid convergent descent method for minimization. Computer J. 6, 163–168 (1963)
8. Fletcher, R., Powell, M.J.D.: A new approach to variable metric algorithms. Computer J. 13, 317–322 (1970)
9. Fletcher, R.: Practical Methods of Optimization, 2nd edn. John Wiley & Sons, Chichester (1987)
10. Goldfarb, D.: A family of variable metric methods derived by variational means. Mathematics of Computation 23, 23–26 (1970)
11. Lechman, M., Stachurski, A.: Nonlinear Section Model for Analysis of RC Circular Tower Structures Weakened by Openings. Structural Engineering and Mechanics 20, 161–172 (2005)
12. Nowak, Z., Stachurski, A.: Nonlinear Regression Problem of Material Functions Identification for Porous Media Plastic Flow. Engineering Transactions 49, 637–661 (2001)
13. Nowak, Z., Stachurski, A.: Global Optimization in Material Functions Identification for Voided Media Plastic Flow. Computer Assited Mechanics and Engineering Sciences 9, 205–221 (2002)
14. Nowak, Z., Stachurski, A.: Identification of an Augmented Gurson Model Parameters for Plastic Porous Media. Foundations of Civil and Environmental Engineering 2, 171–179 (2002)
15. Nowak, Z., Stachurski, A.: Modelling and identification of voids nucleation and growth effects in porous media plastic flow. Control and Cybernetics 32, 820–849 (2003)
16. Nowak, Z., Stachurski, A.: Robust Identification of an Augmented Gurson Model for Elasto-plastic Porous Media. Archives of Mechanics (Archiwum Mechaniki Stosowanej) 2, 125–154 (2006)
17. Shanno, D.F.: Conditioning of quasi-Newton methods for function minimization. Mathematics of Computation 24, 27–30 (1970)
18. Stachurski, A.: "Superlinear Convergence of Broyden's Bounded Θ-Class of Methods. Mathematical Programming 20, 196–212 (1981)
19. Stachurski, A., Wierzbicki, A.P.: Introduction to Optimization (Podstawy Optymalizacji, in polish). Publishing House of the Warsaw University of Technology, Warszawa (1999)
20. Stachurski, A., Lechman, M.: On Solving a Set of Nonlinear Equations for the Determination of Stresses in RC Ring Sections with Openings. Communications in Applied Analysis 10, 517–536 (2006)

21. Stachurski, A.: Orthogonal Projections in the Quasi-Newton Variable Metric Updates. In: Paper presented at the International Conference on Modelling and Optimization of Structures, Processes and Systems, held in Durban, January 22-24 (2007); to appear in IMACS Journal of Mathematics and Computers in Simulation
22. Stachurski, A.: On the Structure of Variable Metric Updates. International Journal of Pure and Applied Mathematics 4, 469–476 (2009)
23. Stoer, J.: On the Convergence Rate of Imperfect Minimization Algorithms in Broyden's β-class. Mathematical Programming 9, 313–335 (1975)
24. Szyld, D.B.: The many proofs of an identity on the norm of oblique projections. Numerical Algorithms 42, 309–323 (2006)
25. Sun, W., Yuan, Y.-X.: Optimization Theory and Methods. Nonlinear Programming. Springer, Berlin (2006)
26. Xiao, Y., Wei, Z., Wang, Z.: A limited memory BFGS-type method for large-scale unconstrained optimization. Computers and Mathematics with Applications 56, 1001–1009 (2008)

Author Index

Printed in the United States
By Bookmasters